U.S. Marine Corps Combat Water Survival

Plus

Special Forces Survival Evasion and Recovery

Department of Defense

Pentagon
Publishing

MCRP 3-02C

Marine Combat Water Survival

U.S. Marine Corps

PCN 144 000069 00

To Our Readers

Changes: Readers of this publication are encouraged to submit suggestions and changes that will improve it. Recommendations may be sent directly to Commanding General, Marine Corps Combat Development Command, Doctrine Division (C 42), 3300 Russell Road, Suite 318A, Quantico, VA 22134-5021 or by fax to 703-784-2917 (DSN 278-2917) or by E-mail to **morgann@mccdc.usmc.mil**. Recommendations should include the following information:

- Location of change
  - Publication number and title
  - Current page number
  - Paragraph number (if applicable)
  - Line number
  - Figure or table number (if applicable)
- Nature of change
  - Add, delete
  - Proposed new text, preferably double-spaced and typewritten
- Justification and/or source of change

Additional copies: A printed copy of this publication may be obtained from Marine Corps Logistics Base, Albany, GA 31704-5001, by following the instructions in MCBul 5600, *Marine Corps Doctrinal Publications Status*. An electronic copy may be obtained from the Doctrine Division, MCCDC, world wide web home page which is found at the following universal reference locator: **http://www.doctrine.usmc.mil**.

Unless otherwise stated, whenever the masculine gender is used, both men and women are included.

DEPARTMENT OF THE NAVY
Headquarters United States Marine Corps
Washington, D.C. 20380-1775

6 January 2003

FOREWORD

Marine Corps Reference Publication (MCRP) 3-02C, *Marine Combat Water Survival*, provides Marine Corps combat water survival techniques, procedures, and training standards. This publication also teaches Marines to cross water obstacles and perform water rescues correctly and safely.

This publication is the foundation for teaching Marines correct water survival techniques and procedures that are used throughout the Marine combat water survival program (MCWSP). Once an individual or a unit has completed the MCWSP, this publication can be used as a refresher course before water operations.

MCRP 3-02C supersedes Fleet Marine Force Reference Publication (FMFRP) 0-13, *Marine Combat Water Survival*, dated 16 September 1991.

Reviewed and approved this date.

BY DIRECTION OF THE COMMANDANT OF THE MARINE CORPS

EDWARD HANLON, JR.
Lieutenant General, U.S. Marine Corps
Commanding General
Marine Corps Combat Development Command

Publication Control Number: 144 000069 00

MCRP 3-02C Marine Combat Water Survival

Table of Contents

Chapter 1. Survival at Sea

Chapter 2. Water Rescues

Chapter 3. Treatment of Casualties and Avoidance of Dangerous Marine Life

Chapter 4. Negotiating Water Obstacles

Chapter 5. Fording Waterways

Appendix. Knot Tying

Throughout history, water has posed special challenges to Marines and Sailors during times of both peace and war. Therefore, the inherent dangers associated with waterborne operations demand that Marines and Sailors receive proper water survival training. Combat units that are confident in their ability to work in and around water can use the water to their advantage in combat, and history is filled with examples of how the proper preparation or training for survival in water reduced or averted disaster.

USS *Indianapolis*

On Sunday, 29 July 29 1945, the heavy cruiser USS *Indianapolis* (the *Indy*) was en route to the Philippine Sea. Shortly before midnight (about 39 hours out of port), the *Indy*, running blacked out and unescorted, was rocked by two explosions on her starboard side. With communications smashed, the ship could not signal its distress and sank within 15 minutes.

Three life rafts and a floater net supported a few survivors, but the rest of the survivors drifted about, held up by rubber life belts or Mae Wests. About 60 seamen died the first night. Survivors assumed the ship would be reported overdue in Leyte, and that they would be rescued within 2 days. By Monday evening, however, panic began to set in as some life jackets lost their buoyancy from the long immersion. Some men even fought over life jackets, which resulted in at least 25 deaths. No one dared sleep for fear of losing his jacket.

Throughout the next several days, in-transit aircraft flew nearby without spotting the desperate seamen. As best they could, the men kept together, some tying long ropes to each other, floating like corks on a net.

Not until late Thursday morning, 3 1/2 days after the ship sank, were the men discovered. Luckily, a plane on a routine flight over the area spotted the survivors. When ships picked them up that night, the survivors learned they had never been reported overdue. Every one of the 1,196-man crew was a casualty; 880 crewmembers were listed as dead or missing. Although many lives were lost, the innovative and expedient use of flotation devices and float techniques employed by the survivors helped save hundreds of lives.

USS *America*

On Thanksgiving eve, 23 November 1995, the USS *America* made its way through the Arabian Sea. Twenty-year-old, Marine Lance Corporal (LCpl) Zachary Mayo was unable to sleep and, wanting some fresh air, made his way onto an open-air platform near the aircraft hangar bay, which was three levels below the sleeping quarters. While he was on the platform, the ship veered suddenly, throwing LCpl Mayo through the platform's protective bars and into the sea, 30 feet below.

Frantic, LCpl Mayo called out in vain to the watchmen on the flight deck, which was 64 feet above him. It soon became clear to him that the USS *America* would keep its course into the Gulf of Oman until his absence was discovered at morning muster. The LCpl took a moment to consider his situation. Since land was at least 100 miles away, swimming was suicide; he would have to stay afloat until a search party found him.

Using the techniques he had learned during combat water survival training, LCpl Mayo made a flotation device out of his coveralls and tried to relax. Meanwhile, business continued as usual aboard the USS *America.* Since LCpl Mayo was on special assignment with the hazardous materials division, his absence went unnoticed until a petty officer asked several Sailors if they had seen their shipmate recently. By the time a roll call had been completed, LCpl Mayo had been adrift at sea for over 24 hours. Although three, fixed-wing Viking aircraft were deployed to search for LCpl Mayo, most people aboard ship feared the worst.

After 34 hours at sea, LCpl Mayo was discovered by fishermen on a Pakistani fishing boat. LCpl Mayo's survival is a testament not only to his incredible physical courage, but also to the soundness of the lifesaving training and techniques he received during combat water survival training.

Chapter 1
Survival at Sea

As a Marine, you face a variety of potential water emergencies whenever you cross expanses of water: ships, watercraft, and amphibious assault vehicles (AAVs) can sink; aircraft can crash into the sea; or you can accidentally fall into the water. However, there are some basic precautionary measures you can take to protect your safety and reduce your chances of becoming a water casualty. Determine the following information as soon as you board any type of vessel. Your knowing the following information may save your life or the life of your fellow shipmates.

- How many life preservers and lifeboats/rafts are on board?
- Where are the life preservers and lifeboats/rafts located?
- What type of unit survival equipment is on board?
- Are individual survival kits issued to each person on board?
- How much food, water, and medicine do the survival kits contain? When was the last time the contents were inspected for proper quantities and shelf life expiration?
- Is there sufficient survival equipment available for the number of personnel?
- How many other personnel are there on board, and where are they located?
- What are the egress procedures for the ship, boat, watercraft, AAV, or aircraft?

Abandoning Ship

When you embark on a Navy ship, you will receive abandoning ship instructions from Navy personnel. If given the order to abandon ship, report to your designated assembly area and put on a life preserver. DO NOT inflate the life preserver until you are clear of the ship. Torn life preservers will not inflate and inflated life preservers can block you, and those behind you, from exiting the ship. A flotation device that has been inflated may also burst if you jump from a significant height. See pages 1-15 through 1-35 for staying afloat with and without a life preserver.

DO NOT remove your clothing, boots, or shoes before abandoning ship. Your trousers and blouse may be the only flotation devices available if your life preserver is faulty or becomes damaged, and your clothes can provide some insulation from the cold water. However, remove your soft cover and place it in a cargo pocket for later use. The soft cover is both lightweight and a good protection against sunburn caused by the sun's rays reflecting off the water.

Jettisoning Equipment

Equipment should be kept properly packed and waterproofed in case you have to abandon ship. If entering the water from a height greater than 30 feet, wearing your equipment (e.g., pack, helmet, gas mask) could cause injury. Upon impact with the water, the helmet will "cup" air inside of it. The chin strap may also create a "hanging effect" as you submerge from the force of the fall. Therefore, you should remove your helmet and gas mask before abandoning ship.

If you are unable to maintain buoyancy due to the amount of equipment secured to your pack and body, then jettisoning some of your equipment may become necessary. Equipment that you should always retain includes canteens of freshwater, first aid kit, soft cover, and survival kit. The survival kit should include first aid items, water purification tablets or drops, fire starting equipment, signalling items (e.g., flashlight, strobe light, chemlights), food procurement items, and shelter items. Other items in a survival kit include sunburn lotion and lip balm, knife, goggles/sunglasses, plastic bag, matches and lighter, mirror. See Marine Corps Reference Publication (MCRP) 3-02F, *Survival*, for a detailed list of survival items and applications.

Abandoning Ship Technique

When abandoning ship, safety considerations must be observed. Use the following technique when abandoning ship without your combat gear:

Place your hands on their opposite shoulders, forming a criss-cross pattern.

Step to the edge of the ship's deck and check the water below for debris or survivors. If the water is clear, look straight ahead and prepare to jump. If the water is not clear, move to another location.

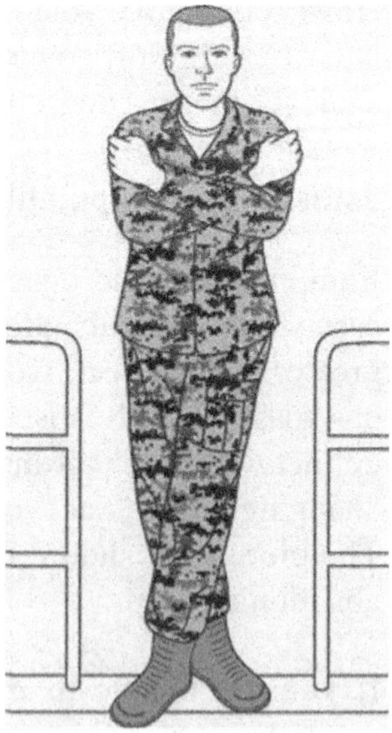

NOTE: DO NOT hold your nose as you abandon ship. If you do hold your nose, the force of impact into the water could jar your arm and hand and cause you to break your nose.

Step off the side of the ship with a smooth, 30-inch stride. DO NOT DIVE OFF THE SHIP. DO NOT LOOK DOWN AT THE WATER. LOOK STRAIGHT AHEAD. Looking down at the water can render you unconscious or cause injuries upon impact.

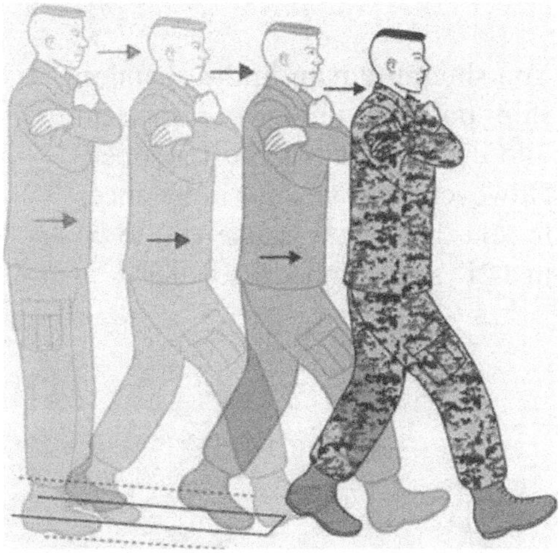

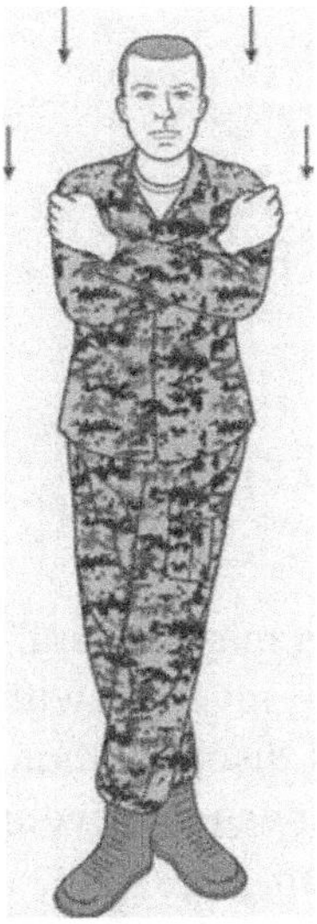

Bring your trailing leg forward during the fall. Cross your trailing leg behind your leading leg.

Keep your head parallel to the water's surface until hitting the water.

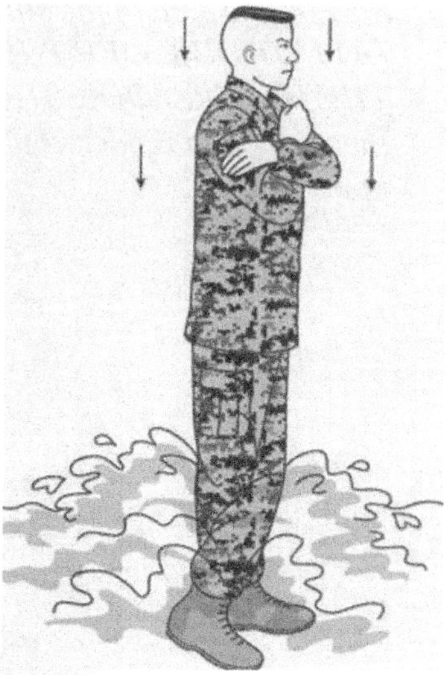

You should remain in the abandon ship position until your descent into the water has almost stopped. However, the weight imbalances in your body may cause you to be in a "J" shape under the water

.

Once your downward motion has ceased, your feet may be parallel with the ocean bottom or you may be nearly inverted with your feet over your head. To counteract potential disorientation, you should pause briefly and allow the natural buoyancy of your torso to bring your body to a nearly upright position.

Floating debris can cause hazards. Therefore, you should swim upward, extending one arm (hand is shaped as a fist) upward to feel for obstructions. If you encounter debris, try to push it away or surface in a different location.

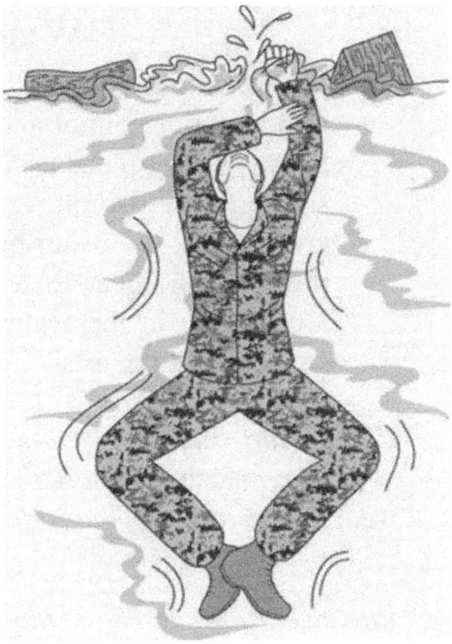

Swim away from the ship. DO NOT LOOK BACK AT THE SHIP. Looking back could slow your movement away from the area. Remember, your objective is to leave the area as quickly as possible because—

- Equipment and debris may be falling from or spilling out of the ship.
- Additional casualties can occur if individuals abandoning the ship fall on top of swimmers already in the water.
- Swimmers close to the sinking ship may get pulled underneath the water by the suctioning effect of the ship as it goes under.

Modified Abandoning Ship Technique

When abandoning ship while wearing full combat gear (weapon, helmet, and a properly waterproofed pack), safety considerations must be observed. The modified abandoning ship technique is used while wearing full combat gear and exiting from a height less than 30 feet.

WARNING

When exiting from a height greater than 30 feet, remove your helmet and pack. Fasten the helmet to the pack or place it inside the pack before jettisoning. If jettisoning gear from the ship or aircraft, check the water below for survivors before throwing the gear forward of the intended jump area. Once in the water, retrieve your gear and swim out of the area.

Place your weapon over one shoulder, muzzle down, with the weapon parallel to your side.

Place your arm and hand along the weapon and hold it to your side.

Take your free hand and place it on top of your helmet to prevent neck/spinal injury from the force of the water pulling upward on the helmet as your body enters the water.

NOTE: An alternate method is to place your weapon over one shoulder, muzzle down, with the weapon parallel to your side. Reach across your body and grasp the sling of your weapon and hold it to your body. Take your free hand and place it on top of your helmet.

Step to the edge of the platform and check the water below for debris and survivors. If the water is clear, look straight ahead and prepare to jump. If the water is not clear, move to another location.

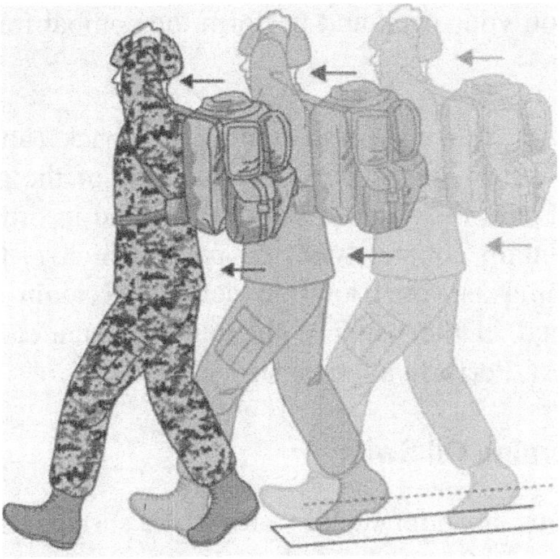

Step off the side of the platform with a smooth, 30-inch stride. DO NOT DIVE OFF THE SHIP, DO NOT LOOK DOWN AT THE WATER, LOOK STRAIGHT AHEAD. Looking down at the water can render you unconscious or cause injury upon impact.

Bring your trailing leg forward during the fall. Cross your trailing leg behind your leading leg.

Keep your head parallel to the water's surface until hitting the water.

You should remain in the modified abandoning ship position until your descent into the water has almost stopped. The buoyancy of a properly waterproofed pack will immediately pull you to the surface. Once you break the water's surface, unsling your weapon and loop the sling over your head. You may have to seesaw the sling so as to ensure the sling passes between your helmet and pack. Once over your head, loop the sling around your neck with the weapon aligned in the center of your body, muzzle down. Lean back on your pack and perform the combat travel stroke to exit the area.

> *Note:* You may also remove the pack (unsnap the quick-release strap on one side of the pack, and pass the strap to your other hand in order to maintain contact with the pack, which is functioning as your flotation device). Remain horizontal in the water with your pack under your chest. Perform a breast stroke kick.

Surface Burning Oil Swim

After you have abandoned ship, rise to the surface using the techniques shown on the page 1-6. However, you must remember that fuel from sinking ships or downed aircraft will float on the surface of the water. Therefore, you must move clear of the floating fuel by swimming away from the ship or aircraft as soon as possible. Either swim upwind (into the wind) of the ship/aircraft or swim against the current. Either method allows you to move away from the fuel and the wind/current will push the fuel past you. To properly execute a surface burning oil swim—

Extend your arms overhead as far as possible.

Wave your arms back and forth vigorously to splash a hole while moving upward.

Splash as long as possible to push burning fuel away from the surfacing area.

Use your arms and hands to sweep away fuel and debris.

Kick your legs in a constant breast stroke kick.

Extend your arms (palms outward) forward on the surface, arms shoulder-width apart.

Pull your hands in and back toward the chest.

Stop your hands in front of your face and rotate them so that your palms face forward (roughly halfway out of the water).

Sweep your arms forward to a full extension at the shoulder width. This splashes debris, oil, or burning liquids aside. To reduce the chance of fatigue, use two short splashes to the front to extend the path.

Repeat the preceding step as necessary while swimming clear of the area.

Surviving With a Pack

If packed properly, your pack will float, and it is your key piece of equipment for staying afloat and overcoming water obstacles. If the pack's contents are properly waterproofed, it can support you (with a combat load) in the water. Buoyed up by a waterproofed pack, you eventually emerge from the water with all your equipment (e.g., boots, helmet, flak jacket, weapon, survival items).

Your pack floats based on a scientific principle known as Archimedes' principle. This principle states that an object submerged in a liquid is buoyed up by a force equal to the weight of the liquid displaced (pushed aside) by the object. If the weight of the displaced liquid is greater than the weight of the object, the object floats. If the weight of the displaced liquid is less than the weight of the object, the object sinks. For example, a machine gun sinks in the water, but it still weighs less in the water than it does on land. Even though a machine gun sinks, it is still buoyed up by a force equal to the weight of the water it displaces. For this reason, you should not try to hold yourself or your equipment any higher out of the water than they would naturally float; doing so wastes both energy and body heat.

Preparing Equipment

Before packing your pack, you must prepare your gear/equipment. Tape or pad all sharp edges and equipment corners. Ideally, your gear/equipment is placed in plastic bags and the plastic bag is then placed inside the standard-issued, rubberized, waterproof bag. Waterproof bags are not completely water tight; henceforth, the added protection of first wrapping the gear/equipment in a plastic bag, then placing the plastic bag inside the waterproof bag. Large plastic bags (e.g., trash bags) work well for bulky equipment (e.g., sleeping bags, field jackets, shelter halves, gas masks). Small plastic bags work best for small items (e.g., shaving gear).

> *Note:* If a gas mask must be carried outside the pack, cover it with a waterproof bag.

You will need the following items, which are available through the supply system, to prepare your equipment for packing:

- Waterproof bags:
  Pistol bag, plastic size #1 (8" x 18").
  Rifle bag, plastic size #2 (10" x 56").
  Machine gun bag, plastic size #3 (15" x 56").
  Multipurpose bag, cover, plastic size #4 (20" x 84").

- Panama work vest (life preserver).
- Small plastic bag, self sealing (6" x 6" and 12" x 12").
- Riggers tape, olive drab (2 1/2").
- Willie peter bag, waterproof clothing.

Tying Waterproof/Plastic Bags

Try to remove the excess air from the waterproof/plastic bag before securing the bag's opening. If filled with air, the bag can burst if pressed from the outside. Perform the following steps to tie a waterproof/plastic bag:

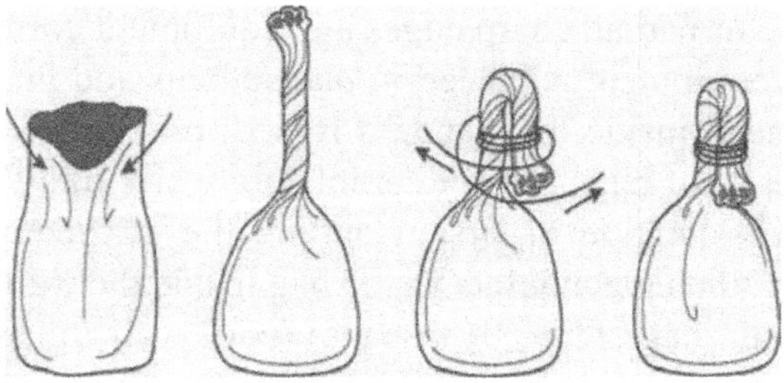

Packing the Pack

Place filled and tied bags inside the pack. Carefully handle sharp items (e.g., tent pegs, poles) to prevent puncturing the bag. Place items in the pack in order of expected use. Close the pack and its compartments. Attach sleeping mats or bags as high as possible on the outside of the pack.

Swimming With the Pack

A useful technique for propelling yourself forward while wearing a pack is the combat travel stroke. This technique allows you to float nearly horizontal and to propel yourself forward with bicycle-style kicks and breast stroke-style arm sweeps. To execute the combat travel stroke—

Body position
Keep the upper part of your body prone to the water, let your legs dangle horizontally, and keep your face up.

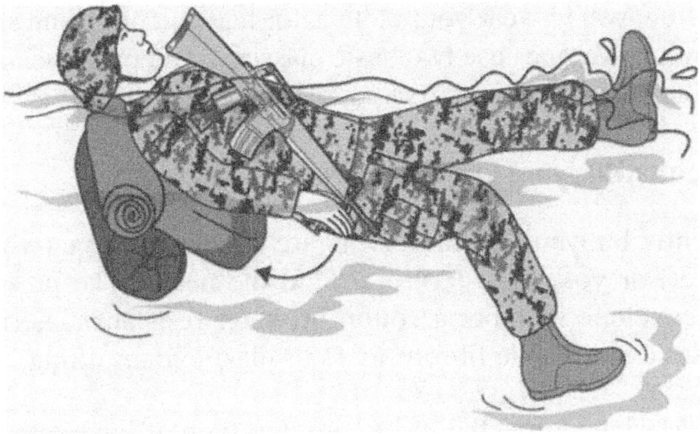

Arm action
Extend your hands out in front of your waist. Sweep your arms slightly downward and back 90 degrees to propel your body through the water. Move your hands back to the front of your waist. Repeat.

Leg action
Continuously move your legs in a bicycle-like movement, bringing your knees up high and step out.

Breathing
Keep your face out of the water during the stroke, and breathe freely.

Staying Afloat With a Life Preserver

The best form of flotation is to find any kind of floating object that will keep you and your equipment out of the water or minimize your exposure to the water. Life preservers are the best method, as they allow you to wear your clothes for heat retention and sunburn prevention. Marines use two basic classes of life preservers: inherently buoyant life preservers and inflatable life preservers.

Inherently Buoyant Life Preservers

Inherently buoyant life preservers are either vest-type (worn like a jacket) or yoke-type (worn around the neck). The preserver's outer envelope is either a cotton or water resistant material that encloses a removable fibrous glass or plastic foam filling.

In the Marine Corps, the most common type of inherently buoyant life preserver is the vest-type with collar, known as the kapock preserver. The kapock consists of collar straps, upper front chest straps, leg straps, and waist drawstrings that secure the preserver to you. The leg straps, which are fitted on both sides of the life preserver, ensure that the preserver remains around your chest while you are in the water. A chest strap is attached to the life preserver to facilitate lifting you out of the water. The strap can also be attached to other survivors or to lifeboats to reduce the fatigue that results from holding onto a floating/secured object by hand.

Inflatable Life Preservers

Marine Corps aircraft and AAVs have inflatable life preservers on board. The inflatable life preserver issued to Marines is known as the LPP (life preserver personal). The LPP is capable of both oral inflation and CO_2 cartridge inflation. The LPP consists of buoyancy chambers, CO_2 inflator, and an oral inflation tube. The buoyancy chamber is deck gray in color and is made from a neoprene-coated nylon fabric.

Inflatable life preservers must be stored in a cool, dry place. Heat, moisture, and light cause deterioration of the life preserver material. Do not stow CO_2 cylinders near steam lines or radiators. Heat can increase the pressure inside the cylinders causing them to explode. Avoid sharp edges in stowage. Sharp edges increase wear and tear on the life preservers and may also puncture inflatable buoyancy chambers.

> **CAUTION:** Do not inflate the life preserver until you are clear of the aircraft, ship, or vehicle. Torn life preservers will not inflate and inflated life preservers can block you, and those behind you, from exiting the aircraft, ship, or vehicle.

To don and adjust the LPP—

Remove the LPP from its storage container.

Fasten the belt fasteners in front with the pouch in the rear.

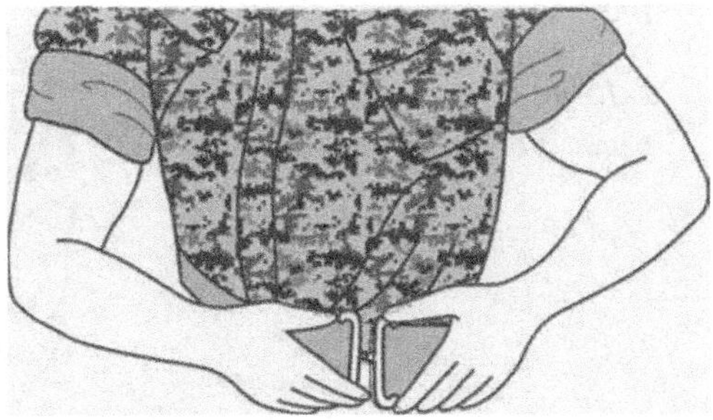

Adjust the belt to fit; secure any excess belt by mating the hook and pile tape.

Rotate the pouch to the front and re-adjust the belt if necessary.

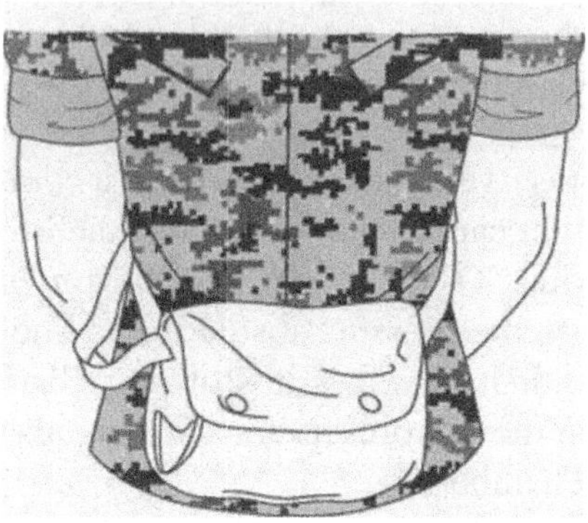

Open the snap fasteners on the pouch and unfold the life preserver.

Place the deflated life preserver over your head.

Place the storage container into the pouch after donning the life preserver.

Lift the lower end of the life preserver out of the pouch.

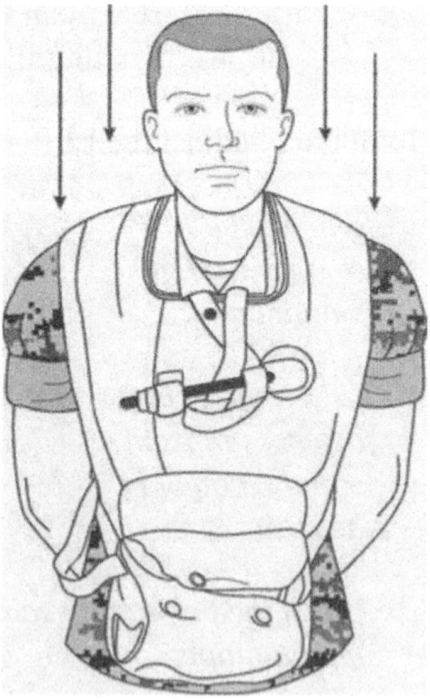

Inflate the life preserver by pulling on the lanyard attached to the CO_2 inflation valve or by blowing on the end of the oral inflation valve.

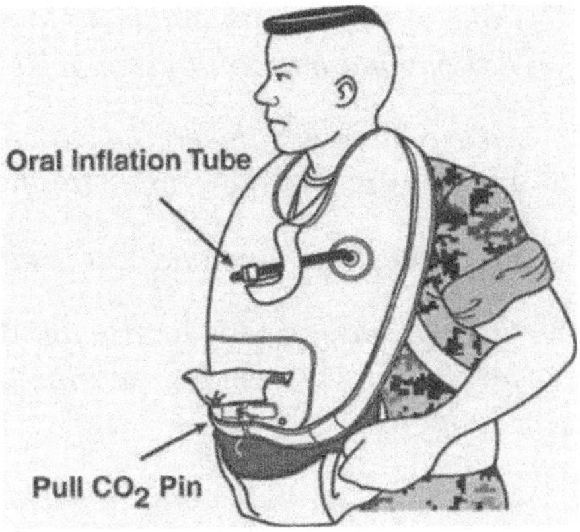

Staying Afloat Without a Life Preserver

You may be in open water without any floating objects or a life preserver to help you survive. If so, uniform blouses and trousers can be made into expedient flotation devices.

Floating With an Inflated Blouse

It is possible to float by a bubble of air trapped in the shoulders of your blouse. The air rises to the back and shoulders of the blouse and supports you at the water's surface. An inflated blouse is also a temporary flotation device used by weaker swimmers while trying to remove their trousers. There is a primary and an alternate way to create a bubble of trapped air in a blouse—

Primary Method
Turn the collar inside the blouse to help create a seal.

Unbutton top button and pull collar around mouth and nose.

Take a deep breath and bend forward slightly at the waist. Exhale one-half to three-quarters of a breath into the blouse.

Grasp and twist the collar with one hand to create a seal, this prevents air from escaping out from the collar.

Use your free hand and feet to stroke and kick to the surface.

Gather and hold the blouse tightly at the collar and stomach level to prevent the blouse from losing air if it floats up too high.

Splash water on the blouse periodically to prevent the material from drying, dry material allows air to escape.

Repeat inflation as required.

Alternate Method
Turn the collar inside the blouse to help create a seal.

Unbutton the second button from the top.

Take a deep breath and bend forward slightly at the waist.

Place your mouth and nose inside the hole created by the open button and exhale one-half to three-quarters of a breath into the blouse.

Grasp material at the unbutton portion and pull downward.

Use your free hand and feet to stroke and kick to the surface.

Splash water on the blouse periodically to prevent the material from drying, dry material allows air to escape.

Repeat inflation as required.

Floating With Inflated Trousers

In warm water, trousers can be used as a primary expedient flotation device. However, in cold water, submerging your head to remove and inflate your trousers results in heat and energy losses that negate the benefit of using the trousers as a flotation device. Once your trousers are inflated, you float motionlessly as if wearing a life preserver. If needed, assume the heat escape lessening posture (known as the HELP position, see page 1-37) to slow heat loss. As trousers dry, air leaks out of the legs. To slow this process, occasionally splash water on the fabric. Reinflate trousers as needed.

Sling Method

The sling method works if you are a strong swimmer or naturally very buoyant. Take the following steps to inflate trousers using the sling method:

Take a deep breath, bend over, and remove your boots.

Note: Retain your boots. Tie the boot laces together and suspend the boots from your blouse or hang them around your neck so that they rest on your chest.

Remove your trousers. Button or zip the trouser's fly closed. This allows you to control airflow.

Tie the bottoms of the trouser legs in a square knot. (The appendix illustrates various knots.)

Ensure that the front (fly) of the trousers faces you.

Hold the trousers above the water's surface and behind your head. Grasp both sides of the waistband and open with both hands.

Kick strongly to stay on top of the water while slinging the trousers overhead in order to trap air into them.

Once the waistband is submerged in the water, air is trapped in the legs.

Hold and seal the waistband underwater.

Slip the inflated legs over your head. Hold the waistband in toward your chest, the fly facing your body. To prevent air from escaping from the trousers, seal the waistband by either folding it or twisting it.

Lie back and relax, resting the back of your neck against the knot.

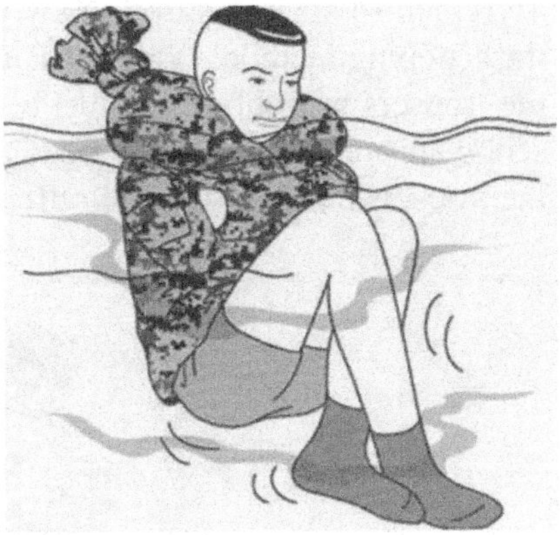

Splash water on the trousers periodically to prevent the material from drying. Dry material allows air to escape.

To replenish air in the trousers, you will use a technique known as the scooping method. With one hand on the open waistband, extend the trousers in front of you just below the surface of the water and scoop air bubbles with your free hand into the open waistband until the trousers have sufficient air. Repeat as necessary.

Splash Method

The splash method is an alternative to the sling method. As with the sling method, you must kick strongly to remain at the surface. To inflate trousers using the splash method, perform the following:

Take a deep breath, bend over, and remove your boots.

NOTE: Retain your boots. Tie the boot laces together and suspend the boots from your blouse or hang them around your neck so that they rest on your chest.

Remove your trousers. Button or zip the trouser fly closed. This allows you to control airflow.

Tie the bottoms of the trouser legs in a square knot. (The appendix illustrates various knots.)

Ensure that the front (fly) of the trousers faces you.

Hold the trousers at the water's surface out in front of you by the waistband with the fly up.

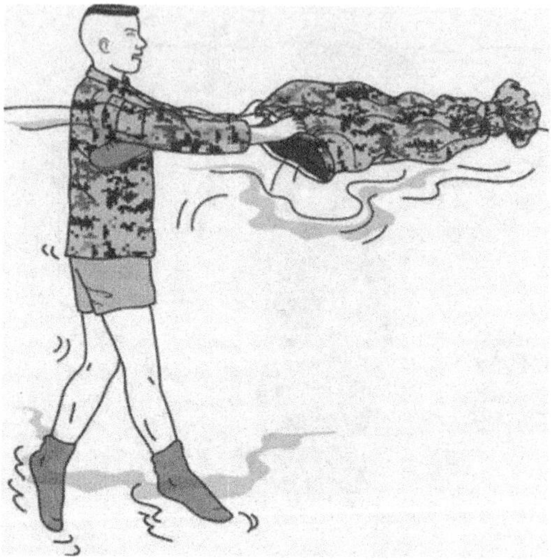

Grasp the waistband at the surface with one hand. Insert your free hand into the waistband, palm down.

Flutter your hand rapidly to create bubbles. This sends a mixture of water and air bubbles into the trousers. The water passes through the fabric. The air remains trapped in the legs.

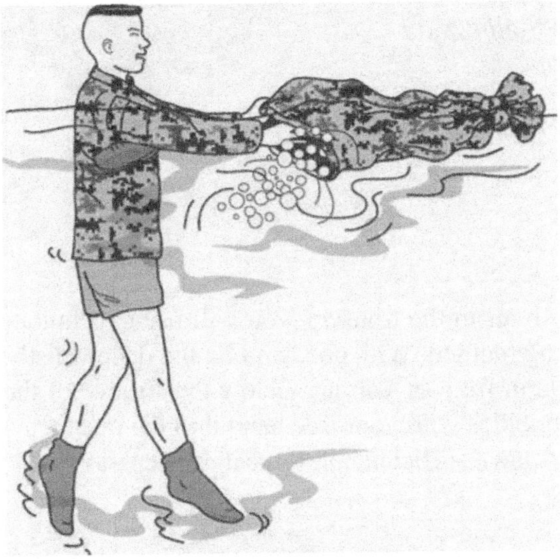

Hold and seal the waistband underwater.

Slip the inflated legs over your head. Hold the waistband in toward your chest, the fly facing your body. To prevent air from escaping from the trousers, seal the waistband by either folding it or twisting it.

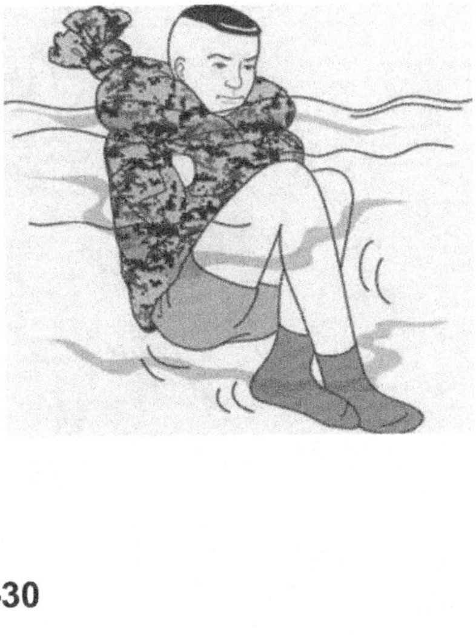

Lie back and relax, resting the back of your neck against the knot.

Splash water on the trousers periodically to prevent the material from drying. Dry material allows air to escape.

To replenish air in the trousers, you will use a technique known as the scooping method. With one hand on the open waistband, extend the trousers in front of you just below the surface of the water and scoop air bubbles with your free hand into the open waistband until the trousers have sufficient air. Repeat as necessary.

Blow Method

The blow method is an alternative to the sling method. Use the blow method if you are a weak swimmer. Take the following steps to inflate trousers using the blow method:

Take a deep breath, bend over, and remove your boots.

NOTE: Retain your boots. Tie the boot laces together and suspend the boots from your blouse or hang them around your neck so that they rest on your chest.

Remove your trousers. Button or zip the trouser fly closed. This allows you to control airflow.

Tie the bottoms of the trouser legs in a square knot. (The appendix illustrates various knots.)

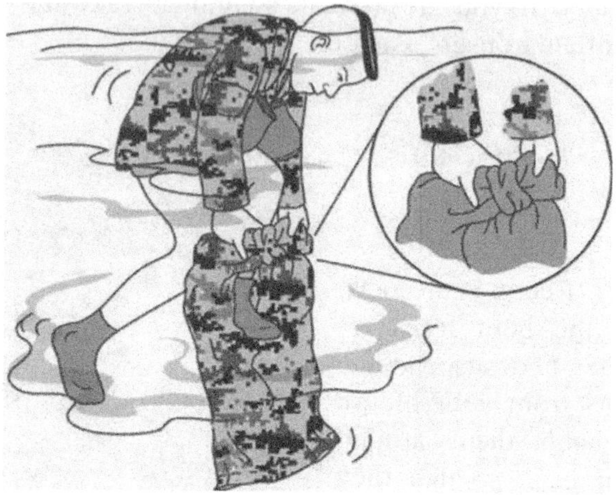

Ensure that the front (fly) of the trousers faces you.

Hold the trousers at the water's surface. Grasp both sides of the waistband and open with both hands.

Take a deep breath.

Drop 2 feet below the water's surface, pulling the waistband underwater.

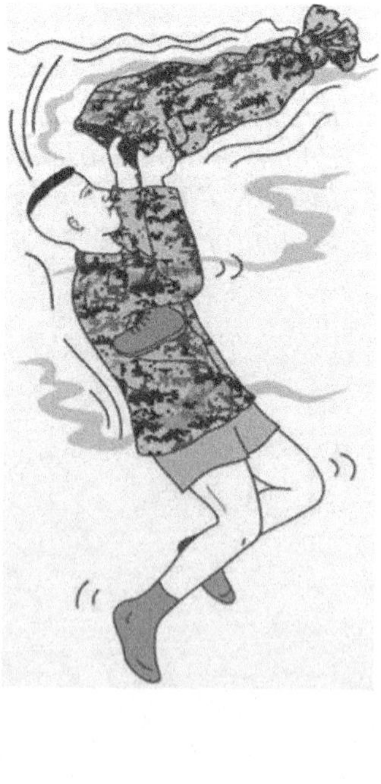

Hold the waistband open with both hands and blow air into the trousers.

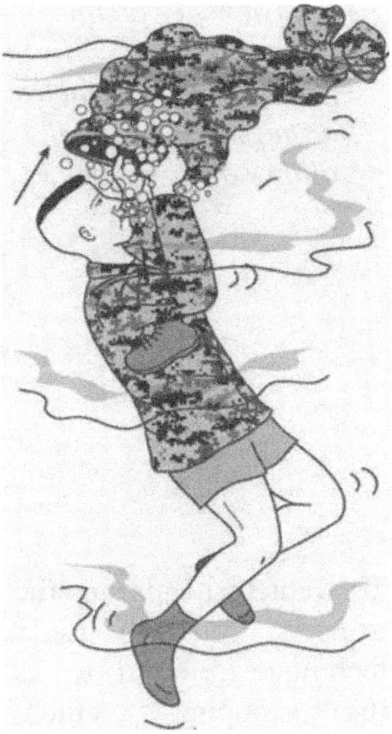

To fill the trousers with air, surface while keeping the waistband underwater, breathe in again, drop below the water's surface, and blow air into the trousers. Repeat these steps until the trousers are filled sufficiently. Once trousers are filled—

Hold the waistband underwater. Twist and pinch it off.

Slip the inflated legs over your head. Hold the waistband in toward your chest, the fly facing your body. To prevent air from escaping from the trousers, seal the waistband by either folding it or twisting it.

Lie back and relax, resting the back of your neck against the knot.

Splash water on the trousers periodically, to prevent the material from drying. Dry material allows air to escape.

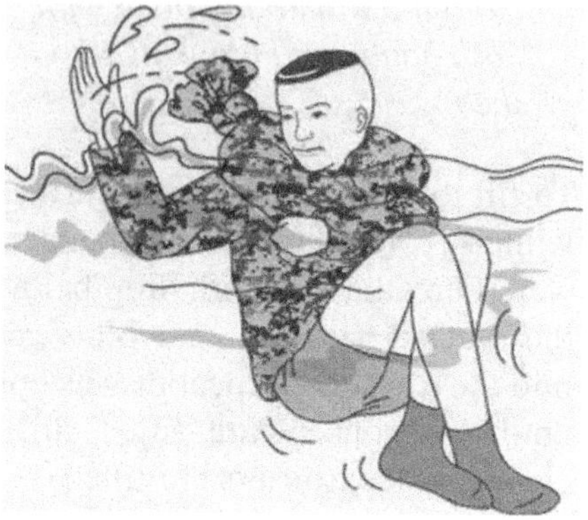

To replenish air in the trousers, you will use a technique referred to as the scooping method. With one hand on the open waistband, extend the trousers in front of you just below the surface of the water and scoop air bubbles with your free hand into the open waist-band until the trousers have sufficient air. Repeat as necessary.

Avoiding Heat Loss in Cold Water

The rate of heat exchange in the water is about 25 times greater than it is in air of the same temperature. When you are immersed in cold water, hypothermia occurs rapidly due to the decreased insulating quality of wet clothing and as a result of water displacing the layer of still air that normally surrounds the body. You also lose about 50 percent of your body heat through your head; therefore, keep your head out of the water. Other areas of high heat loss are the neck, the armpits/sides, and the groin.

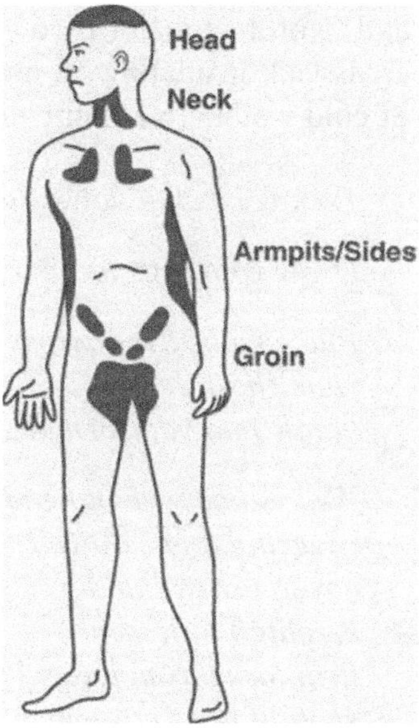

In cold water, DO NOT SWIM TO STAY WARM. Swimming, even with a slow and steady stroke, produces a lot of heat that is lost in the water. The heat loss can produce hypothermia that slows body functions and can result in serious injury or death. Remaining motionless conserves body heat three times longer than swimming. SWIM only if you have a flotation device and the shoreline is visible.

Individual Protection From the Cold

If you are equipped with a life preserver, assume the heat escape lessening posture (known as the HELP position) to slow heat loss and to protect major blood vessels near the body's surface. These areas lack insulating fat and are vulnerable to the chilling effects of cold water. To assume the HELP position—

Tuck your chin down tightly to cover your throat.

Draw your legs up in a fetal position to protect the groin.

Place your arms across your chest, tuck your hands into your armpits.

Wear some type of head covering (e.g., stowed cover, towel, handkerchief) to lessen heat loss through the scalp if head covering is available.

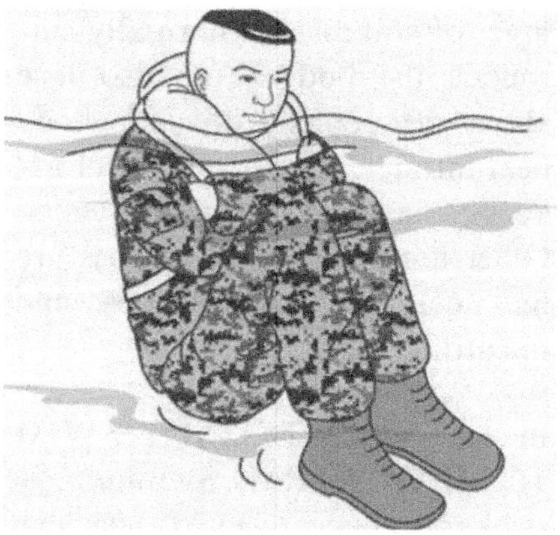

Group Protection From the Cold

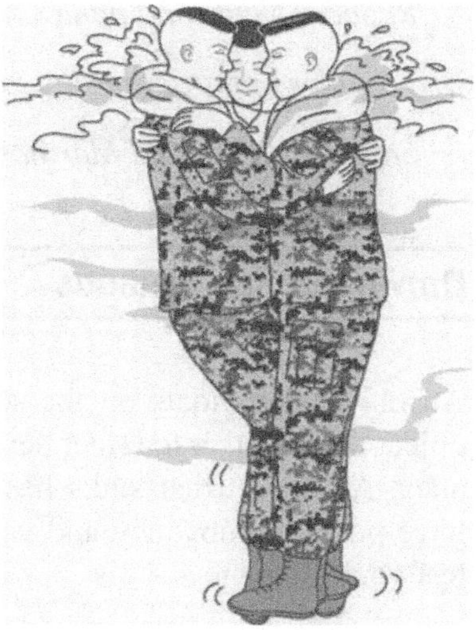

If three or more Marines are in the water and are equipped with life preservers, they should wedge tightly together and lock arms to form a circle known as a huddle position. This position protects vulnerable areas from heat loss.

A casualty who is suffering from the effects of the cold can be placed within the huddle to be surrounded by warmer water. If in the water for a prolonged period, it is recommended that everyone be rotated inside the huddle to maintain or re-warm each person's internal core temperature. If there are more than five Marines, they should make clusters of huddle positions.

Contact with other swimmers provides survival advantages—

Creates a larger target for search and rescue aircraft.

Provides additional warmth in cold water.

Improves morale.

Re-establishes the chain of command.

Reduces shock and panic.

Provides opportunities to administer first aid.

Supports exhausted Marines.

Drownproofing Methods

An object that floats on the surface has positive buoyancy. An object that floats a few feet beneath the surface has neutral buoyancy. An object that sinks has negative buoyancy. Most people have positive buoyancy and will float at the water's surface. To test your buoyancy—

Stand in water that is at your chest level or deeper.

Take a full breath.

Bend forward slowly.

Relax and wait.

If you have positive buoyancy, you will slowly rise to the surface. If you have neutral buoyancy, you will float a few feet beneath the surface. If you have negative buoyancy, you will sink. Regardless of whether you naturally float or sink, you can remain at the surface for extended periods without a life preserver if you exercise the appropriate drownproofing methods or survival strokes, which are based on your buoyancy.

Drownproofing methods consist of the T-method and the sweep. The breast stroke, side stroke, and elementary backstroke are the most common survival strokes. With any drownproofing method or survival stroke, remember the acronym **SAFE**:

SLOW EASY MOVEMENTS
Move slowly to conserve energy and minimize heat loss.

APPLY NATURAL BUOYANCY
Use natural buoyancy to support the body.

FULL LUNG INFLATION
Fill the lungs with each breath. Do not hold air in the cheeks.

EXTREME RELAXATION
Tight muscles are denser than relaxed ones and do not float as well.

Crawl Stroke

The crawl stroke, sometimes called the front crawl or free style, is the fastest stroke. To execute the crawl stroke—

Body position
Lie horizontal, on your stomach, in the water.

Look forward and downward at a 45 degree angle with the waterline between your eyebrows and hairline. Your head position is important as it assists in cutting a path through the

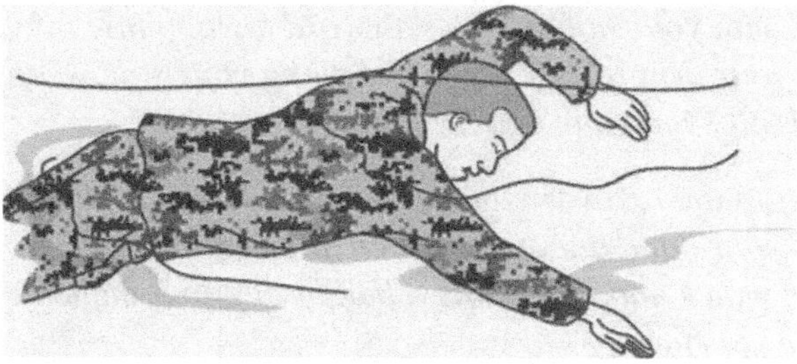

water. If your head is too high in the water, your lower body sinks significantly, making your stroke less efficient. If your head is too low in the water, water washes over your shoulders and neck, causing unnecessary drag.

Arm action

Arm action occurs in three phases: catch, propulsion, and recovery.

Fully extend one arm forward of your body, this positions your hand to catch the water in preparation for the propulsion phase.

To catch the water, bend your wrist (with your palm pointing outboard) and make an "S" shape (or inverted "S" shape) with your hand, ensuring that your hand does not cross the center of your body. Your left hand makes an "S" shape and your right hand makes an inverted "S" shape with your hand finishing at shoulder level.

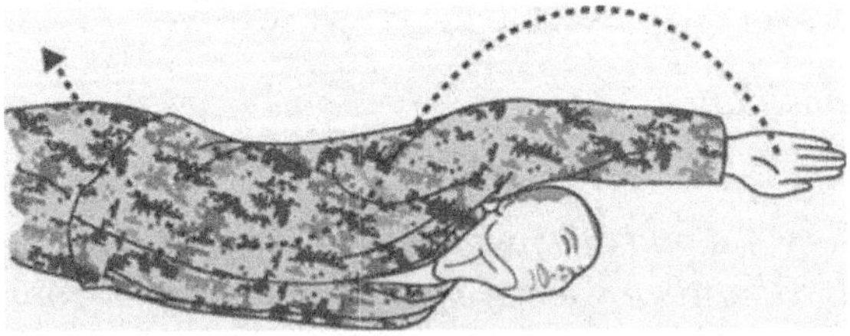

Push with your hands in a rearward fashion toward your feet until your arm is fully extended along your side, keeping your hands close to your body.

To begin the recovery phase, bend your arm at the elbow and raise your hand out of the water. Your hand breaks the surface of the water and maintains a height of 2 to 3 inches above the water's surface.

With your hand and arm moving in a forward manner, bring your hand past your head until your arm is about three-quarters of the way extended. At this point, turn your hand with your palm outboard so as to allow your thumb and fore-finger to enter the water first.

Once your hand enters the water, continue to push your arm forward until it is fully extended.

You are now prepared to catch the water again. These steps are performed in an alternating pattern: when one arm is catching and propelling, the other arm is recovering.

Leg action
Use a flutter kick to create the leg action for the crawl stroke. This kick is used for both propulsion and keeping the lower body horizontal with the water's surface. The flutter kick is an alternating leg action: one leg is kicking in a downward motion while the other leg is recovering to the surface to prepare for the next kick. The size of the kick ranges from 12 to 15 inches and depends on your height.

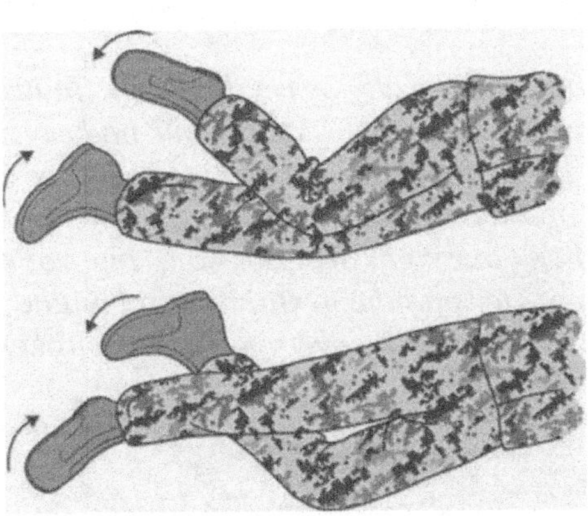

Maintain your legs in a semi-rigid manner.

Generate power for the kick from the hips.

Keep your feet loose at the ankles, they trail behind your legs and act as "flippers."

Execute the propulsion phase of this kick with a downward thrust of your leg.

Execute the recovery phase by pushing your leg back to the surface. This phase is complete when your foot reaches the surface.

Breathing
Breathe during the recovery phase of the arm action.

Roll your body and rotate your head to the side of your body where the arm recovery is occurring, this rolls the water away from your mouth. Keep your chin pushed back toward your shoulder.

Exhale while your face is still submerged and inhale when your face breaks the surface.

Breathe either bilaterally or rhythmically. To breathe rhythmically, breathe on the same side of your body every time your arm cycle occurs. Bilateral breathing requires you to breathe every one and a half arm cycles. To breathe bilaterally, breathe when your right arm is recovering, your face goes back in the water and the next time you breathe is when your left arm is starting its recovery three arm strokes later.

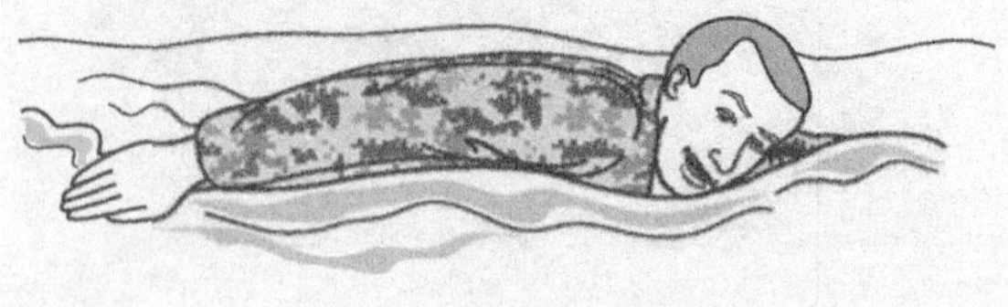

Bilateral breathing is the preferred method for breathing; it prevents the chance of hyperventilation and allows your body to maintain a lateral position in the direction you are swimming.

Coordination
This stroke uses constant arm and leg action.

T-Method

The T-method is a basic drownproofing method. This is the best survival technique if you have negative buoyancy. To execute the T-method—

With your face out of the water, take a deep breath and submerge your face in the water while holding your breath.

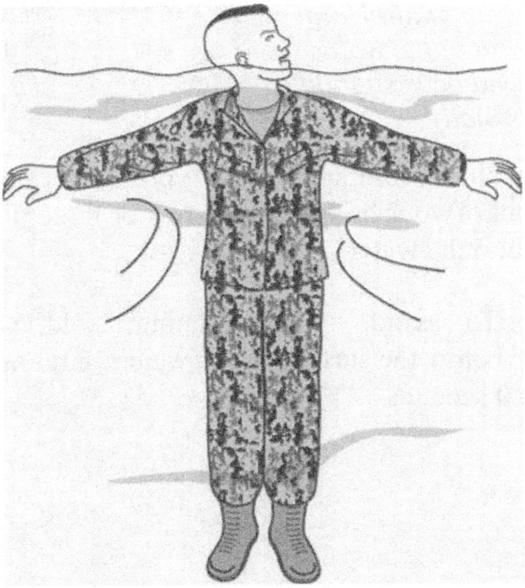

Float with your body in a horizontal position, arms extended from your side, and legs extended and joined.

Move your hands up to your armpits by tracing an imaginary line along your ribs.

Extend your arms outward (horizontal) to your sides, your body position resembles the letter "T."

Step out and forward with one leg and point your other leg to the rear, your knees should be slightly bent. Simultaneously, bring your arms down to your sides. Then bring your legs back together. You complete the steps by exhaling most of your air and preparing to surface your face to obtain another breath of air. Hold your head out of the water. Tilt your head back slightly. Breathe normally.

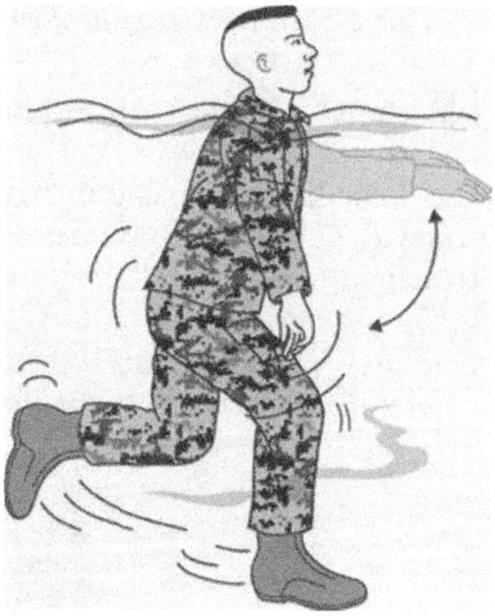

Once your breath is complete, move your hands up and down directly in front of your body. Do this two or three times to slow your descent into the water.

> NOTE: To avoid hyperventilating, hold your breath below the surface of the water for no more than 10 seconds.

The Sweep

The sweep works well if you have slight to excellent positive buoyancy. To execute the sweep—

Float face down in the water: bend 45 degrees at the waist, arms and legs dangling, head hanging down, relax all muscles.

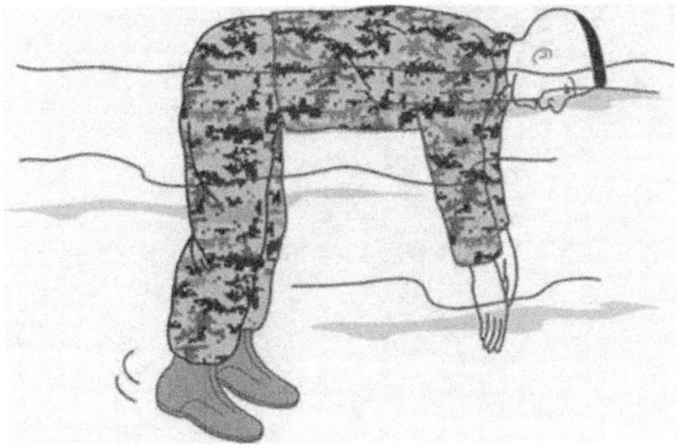

Spread your feet slowly to prepare for a single kick (one leg is forward and one leg is rearward).

Cross your arms in front of your chest, palms outboard with the back of each hand touching the opposite ear.

Exhale prior to raising your head for a breath.

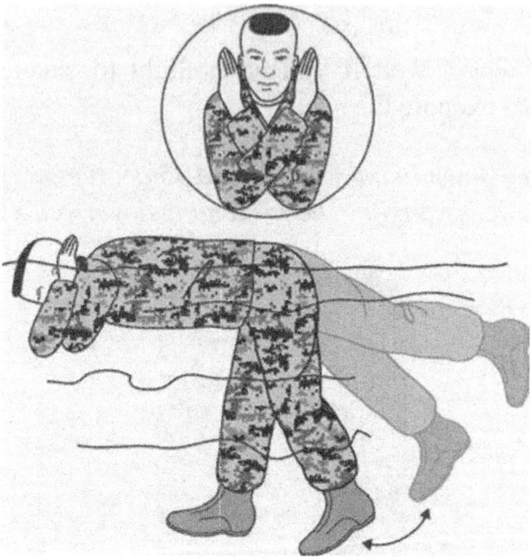

Bring your legs together and sweep your arms down and out until the arms are fully extended out to the sides. This raises your face above the water and allows you to catch a breath of air.

With your air supply replenished, return your face to the water and relax while sweeping your arms in a downward motion in front of your body to prevent/slow your descent. Do not hold your breath for more than 10 seconds. More than 10 seconds enhances your chances of shallow water black out and subsequent drowning.

Breast Stroke

Use this stroke to swim underwater, through oil or debris, and in rough seas. If you are a good swimmer and not wearing combat gear, the breast stroke is the best stroke for long-range swimming because it provides good visibility and allows you to conserve your energy and maintain a reasonable speed.

Body position
Lie prone in the water. Swim with your trunk and legs projecting back and down at an angle of 20 to 30 degrees.

Extend arms out in front (hands together [side by side]), and extend legs behind (toes pointed) to prevent drag.

Face downward, looking forward at a 45 degree angle to break the water and to prevent water from washing into the collar area causing drag.

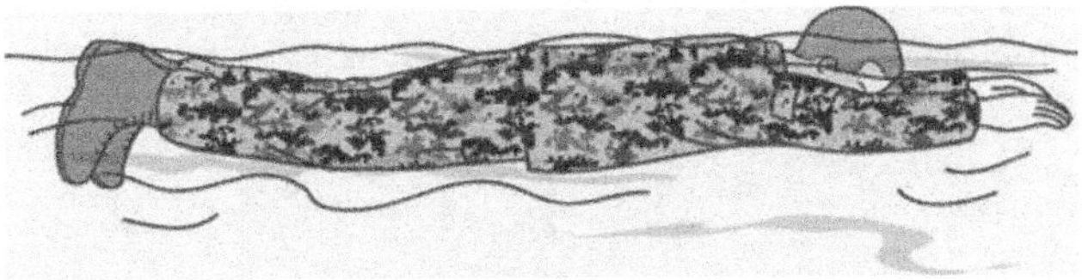

This is known as a glide.

Arm action

Turn your palms outward and bend your arms slightly.

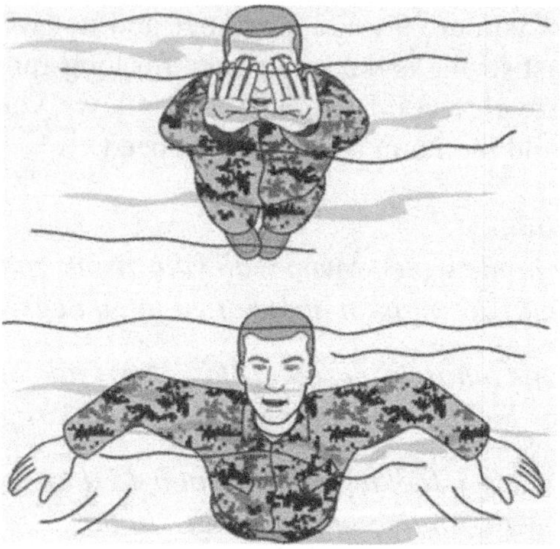

Sweep your arms sideward and slightly downward until your hands are opposite and slightly below your shoulders.

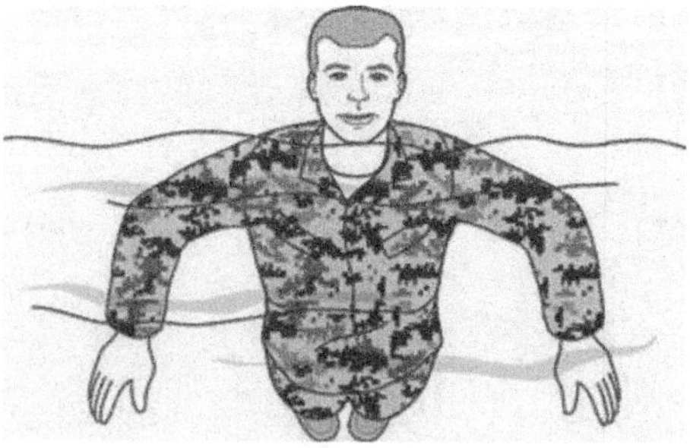

Rotate your head up, breathe once your mouth breaks the surface.

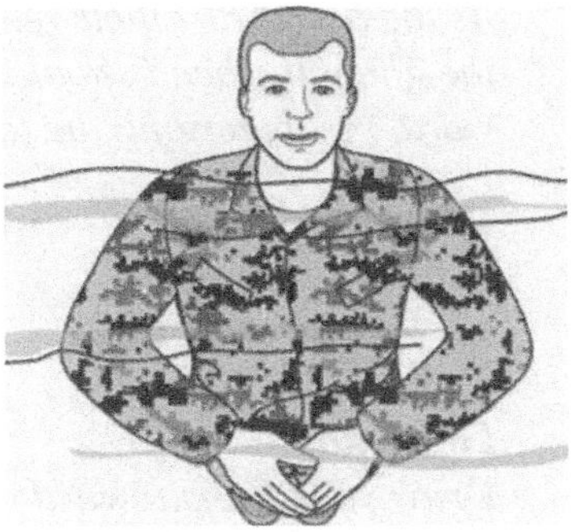

Bring your hands and arms up along your chest and thrust them forward until they are extended and ready to execute the next arm pull.

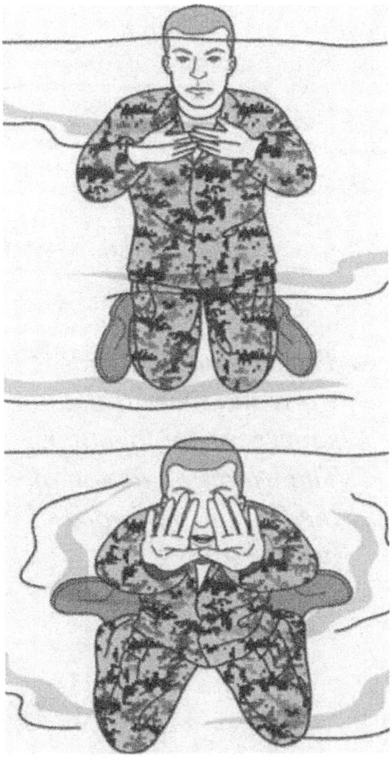

As the arms start their recovery into the glide, the head should rotate forward, resubmerging the face.

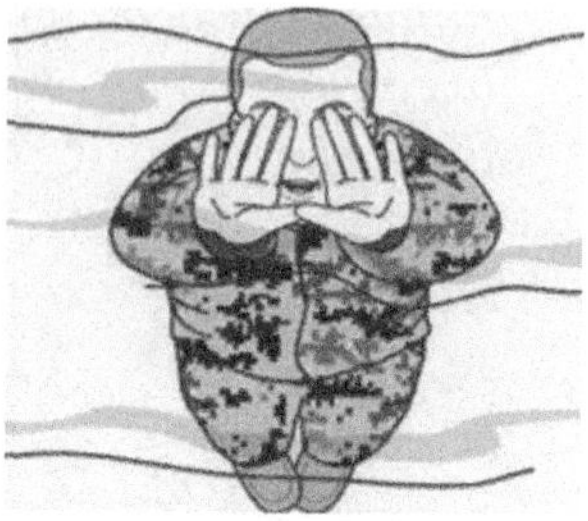

Leg action

Draw your heels toward your buttocks, establish a 45 degree bend in the knees.

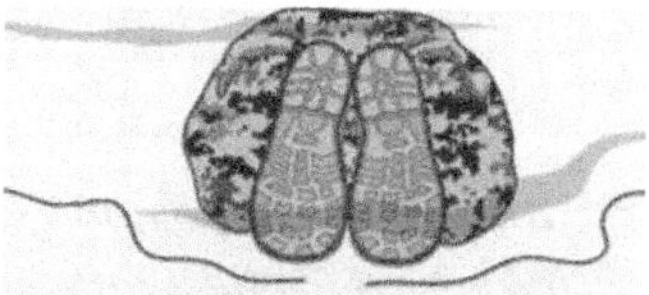

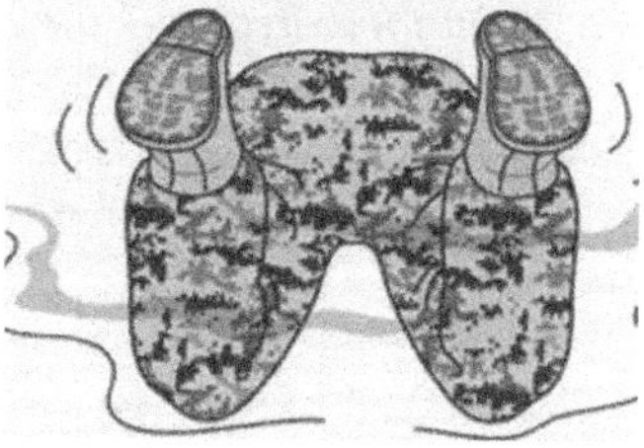

Thrust your legs outward and rearward, then squeeze them together. The whipping action of the feet aids forward propulsion.

This is known as the breast stroke kick.

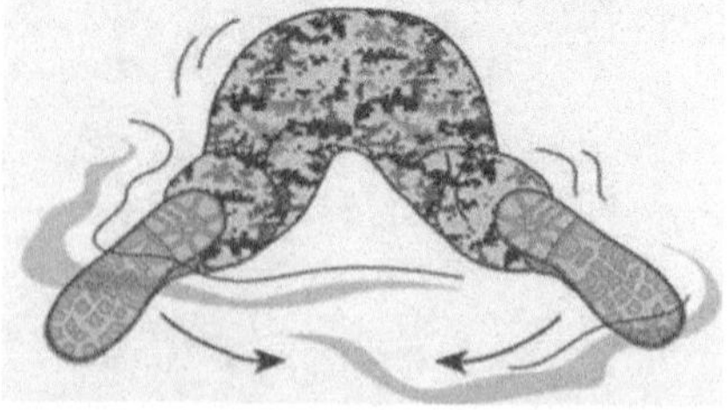

Breathing
Inhale during the arm pull and exhale through your mouth and nose during the finish of the breast stroke kick and glide.

Coordination
The stroke movement is in three counts:

Begin your arm pull. Near the finish of the pull, flex your knees and bring your heels toward your buttocks. The arm pull counteracts the resistance created by the knees.

As the arm pull is completed, thrust your hands forward, kick your legs outward and rearward, and squeeze them together.

Glide through the water for approximately 1 to 3 seconds or until your forward momentum decreases, then begin the next stroke.

Side Stroke

The side stroke is a survival stroke because you use both arms for buoyancy, with each arm creating a slight propulsion. The majority of your body's propulsion comes from your kick. To execute the side stroke—

Body position
Lie on your side with your lead (bottom) arm extended beyond (with a slight bend in your elbow) your head and in line with your body. Palm is down and your hand is submerged 6 to 8 inches.

Extend your trail (top) arm down the length of your body over your thigh.

Keep your legs straight and together, toes pointed rearward.

Keep your face out of the water; this allows for free breathing.

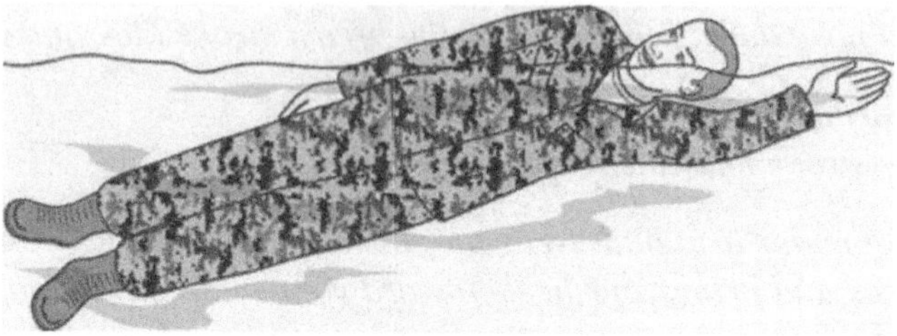

This is known as the glide.

Arm action
With your lead arm, pull your arm downward, while flexing at the elbow, until it is straight down from your shoulder.

Rotate your shoulder and pull your elbow into your side. This should put your lead hand at shoulder level. At the same time, turn your palm toward your face and thrust forward to your original, extended position.

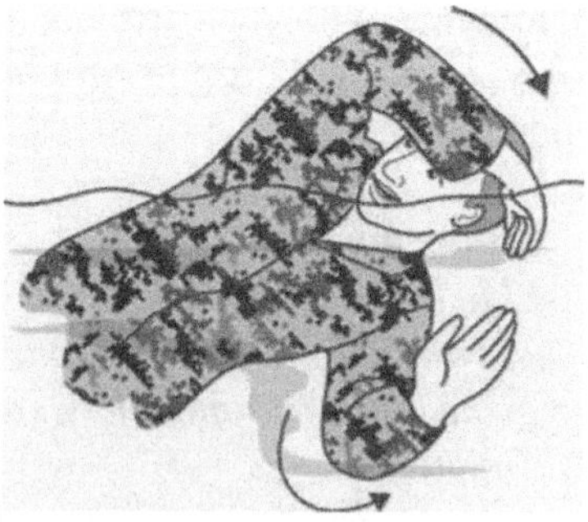

Draw your right hand upward in front of your chest to shoulder level. Rotate your palm toward your feet, then push it downward in front of your body toward your feet to catch the water.

Push your trail hand backward to its original position on top of your thigh. (Your trail hand starts forward and meets your lead hand at your chest/shoulder.)

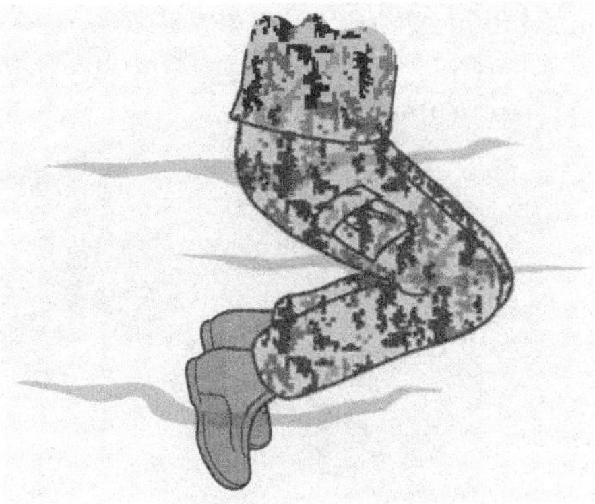

Leg action
To perform the scissor kick, the top leg always goes forward and the bottom leg always goes rearward. From the extended position, draw—or recover—your feet toward your buttocks until your legs are bent at a 45 degree angle at the knees and the hips are flexed at a 45 degree angle with the thighs.

Once the legs have completed their recovery and while maintaining a 45 degree bend in the knees, extend the legs fully into a "V" shape in order to catch the water for the propulsion phase.

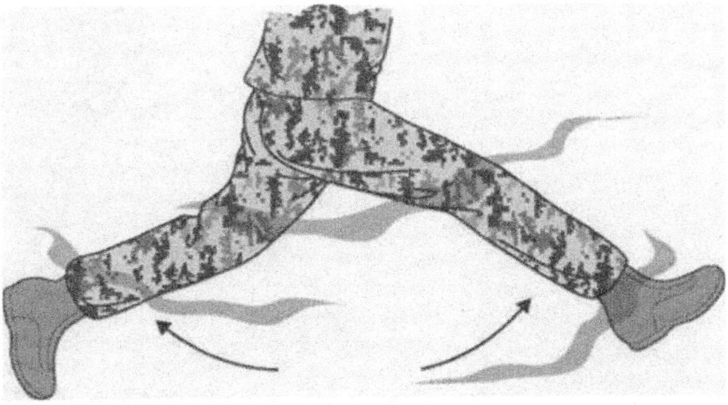

Once the legs are separated and extended forward and rearward to the "V" position, sweep the legs together until the feet are together.

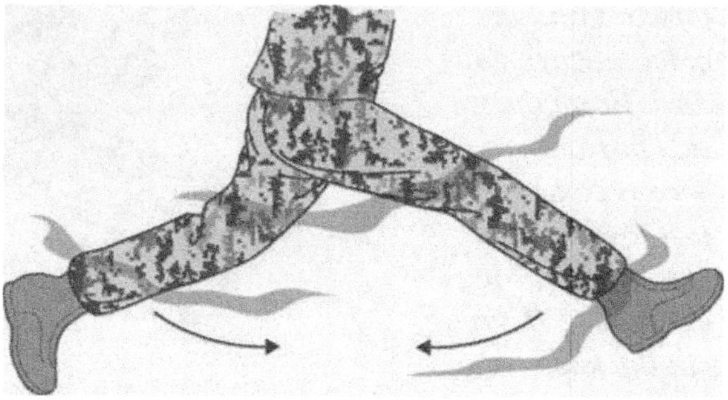

You are now in the glide position.

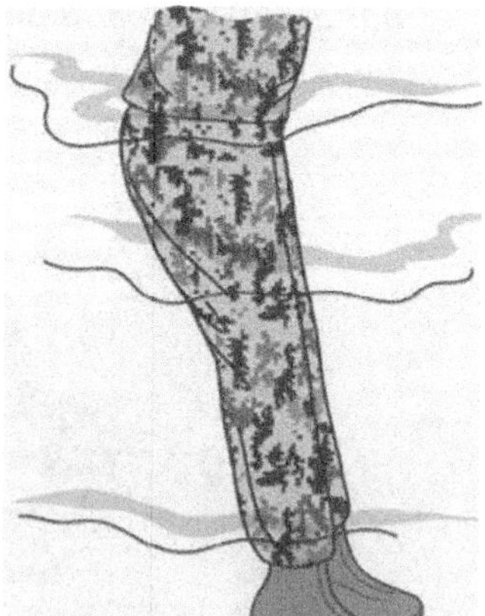

Breathing

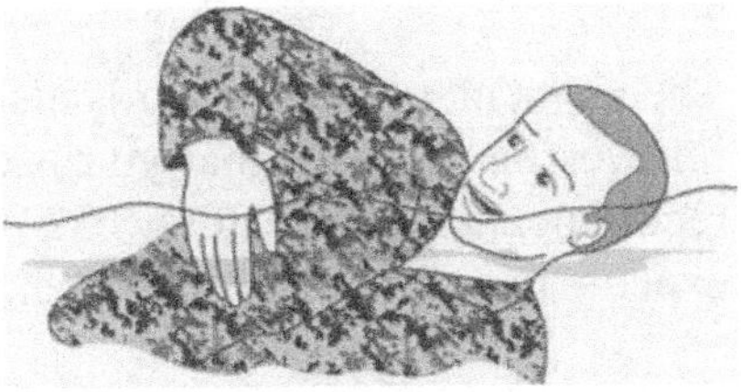

As long as your face remains clear of the water, it is a free breathing stroke. However, it is recommended that you exhale then quickly inhale when the legs are sweeping back together in the scissor kick. This is when the body reaches its highest point in the water, thus clearing the face completely from the surface of the water making it the optimum time to breathe.

Coordination

Begin the stroke with the downward pull of your lead arm. At the same time, bring your trail arm upward and draw your knees up to begin the kick. Let the thrust of the lead arm, push of your trail arm, and the kick of your legs coincide in order to finish the glide position. Glide through the water for approximately 1 to 3 seconds or until your forward momentum decreases, then begin the next stroke.

Elementary Backstroke

The elementary backstroke is also an excellent survival stroke. It relieves the muscles that you use for other strokes, and it is the recommended stroke for weak swimmers or nonswimmers. To execute the elementary backstroke—

Body position
Start on your back.

Face up, chest up, and hips up, keeping an arch in your lower back with arms pressed to your sides and your legs extended and joined to prevent drag.

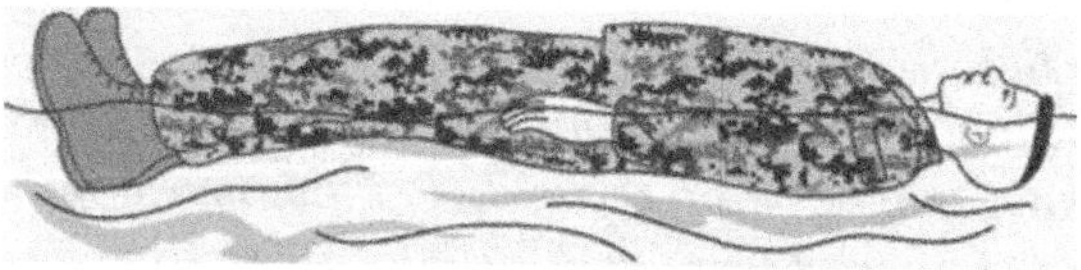

Arm action
Trace your hands up your sides to an area near your armpits then extend your arms out to the sides to form the letter "T" (palms facing feet), locking out the elbows.

> *Note:* Don't raise your arms above your head. This creates drag, changes your body position, and submerges the head.

Slap your palms to your thighs using a strong sweeping motion.

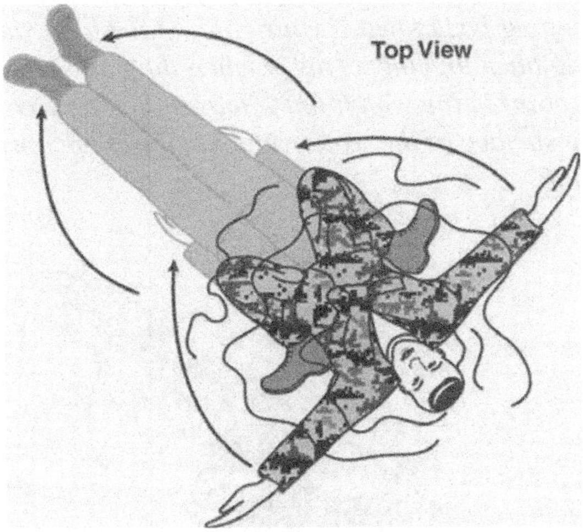

Leg action
Bend both legs at the knee (90 degree angle) slightly separating your knees and drawing your heels downward to a point under and outside your knees. The knees are spread as wide as the hips or slightly wider depending on the body type of the swimmer.

Circle around in a whipping action, ending with legs in a glide position.

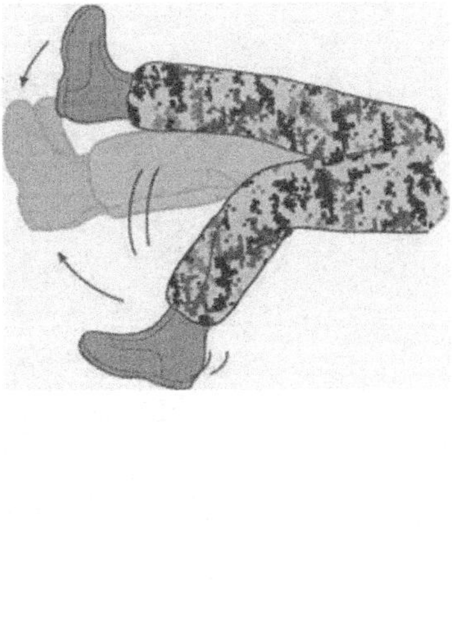

Breathing

Breathe anytime during this stroke. However, it is recommended that you exhale then quickly inhale when your arms are sweeping back toward your sides and while your legs are sweeping back together. This is when the body reaches its highest point in the water, thus clearing the face completely from the surface of the water and making it the optimum time to breathe.

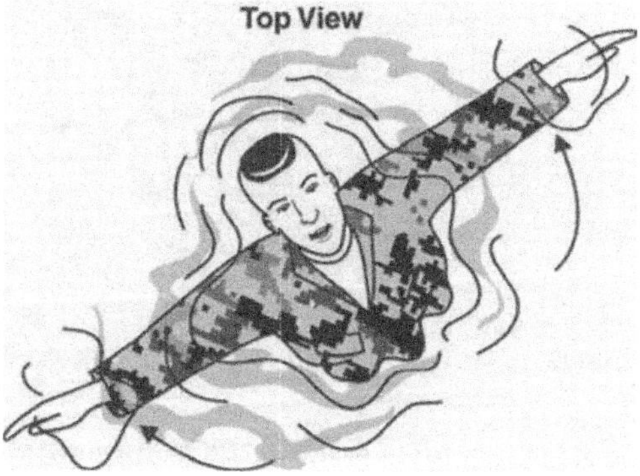

Coordination

The stroke movement occurs in three counts (recovery, catch, power).

Begin the arm pull (recovery).

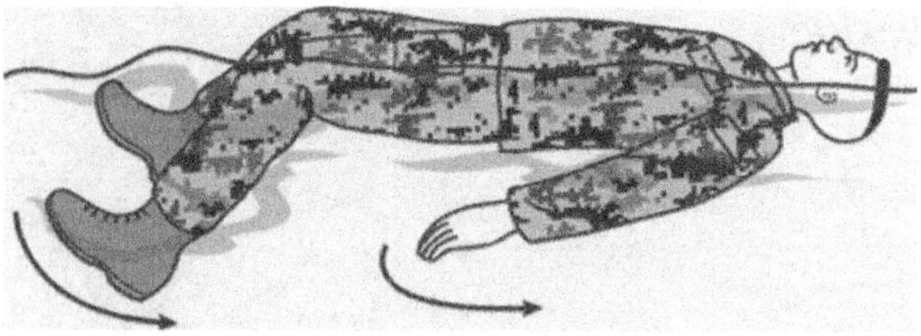

Near the finish of the pull, flex your knees to a 90 degree angle. The arm pull counteracts the resistance created by the knees.

Kick out your legs, and squeeze them together as the arm pull is completed (catch, power).

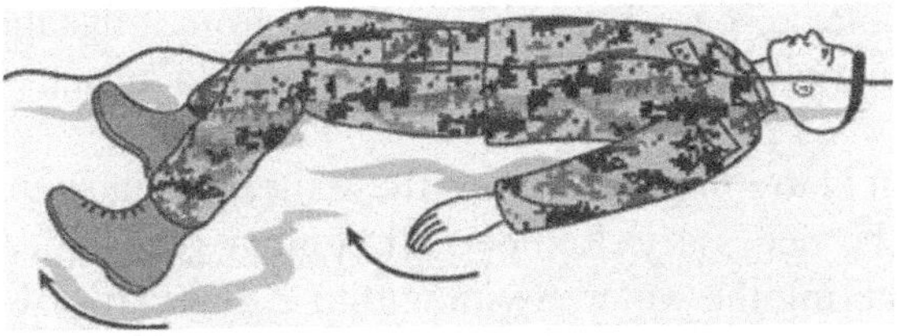

Glide through the water for 1 to 3 seconds or until your forward momentum decreases, then begin the next stroke as your momentum slows.

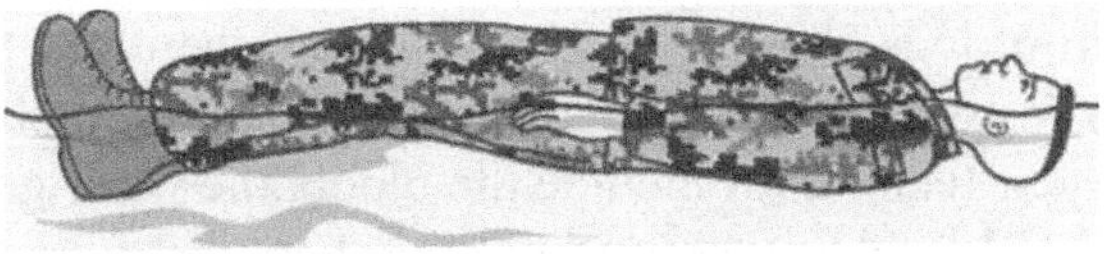

Chapter 2
Water Rescues

A drowning victim can panic and react with unexpected violence and can seize and inadvertently drown a rescuer. Therefore, if possible, a water rescue should be executed from a distance. Reaching, wading, or throwing methods are used in a distance rescue.

If a victim is too far away to use these methods, a swimming rescue may be necessary. Remove all combat gear, when possible, before entering the water. Swim within 2 to 6 yards of the victim to maintain a margin of safety, this allows you to reassess the situation and reassure the victim. If the victim is unconscious, use the wrist tow method or cross-chest carry method to pull the person to safety. If the victim is struggling, use a rear approach and then execute either a single armpit level off or a double armpit level off before towing the victim to safety.

If the victim does begin to overpower you, there are techniques that allow you to defend yourself without having to abandon the rescue. These techniques include the block, the wrist-grip escape, the front head-hold escape, and the rear head-hold escape. These techniques allow you to separate yourself from the victim, reassess the situation, and then attempt the rescue again.

> *Note:* In this chapter's illustrations, the rescuer is shown without a helmet.

Reaching Rescue Techniques

Reach

From a safe position at the water's edge, reach out to the victim. Talk constantly to calm the victim. Retain partial contact with land or some solid support structure (e.g., pier, bridge). If the victim is close but still beyond reach, extend an object (e.g., stick; pack; rifle with magazine removed, chamber empty, and muzzle pointing toward victim) that the victim can grasp. Pull the victim slowly to safety. Once the victim is close to shore, remove the victim from the water.

Note: You can also extend a foot to the victim if you can retain a secure grip on a solid support structure.

Reach From a Deck

Reach from a deck rescue can be executed by either a swimmer or a nonswimmer, and it can be used on an active or a passive victim. Be sure to reassure the victim during the rescue. To execute reach from a deck—

Lie prone on the deck with your body firmly anchored. To anchor your body, lie flat, spread legs apart, and extend one arm behind you with your palm down and on the deck.

Keep as much of your weight on the deck as possible and extend your free hand to the victim.

Grasp the victim's wrist from above, your thumb and index finger are facing you. The victim should never be allowed to grab you and put your life at risk. Therefore, never reach across the victim to grab his wrist, always grab the wrist that is the closest to you.

Keep your arm straight and locked out and pull the victim to the side of the deck. The victim should never be allowed to grab you and put your life at risk. Therefore, never pull the victim into you, always pull the victim into the side.

Arm Extension

An arm extension rescue can be used if you cannot reach the victim using the reach from a deck rescue technique and you must enter the water. An arm extension rescue technique is used for a victim who is either active or passive. Once you determine that a reach from a deck rescue technique is not viable—

Reassure the victim and quickly ease into the water while holding onto the deck with one hand.

Grasp the victim's wrist from above, your thumb and index finger are facing you. The victim should never be allowed to grab you and put your life at risk. Therefore, never reach across the victim to grab his wrist, always grab the wrist that is the closest to you.

Keep your arm straight and locked out and pull the victim to your side. The victim should never be allowed to grab you and put your life at risk. Therefore, never pull the victim into you, always pull the victim into the side.

Leg Extension

If the victim is beyond the reach of your arm, ensure that you have a firm grip on the deck, extend a leg to the victim and allow him to grab it, slowly bring the victim in closer until you can grab the victim's wrist that is holding onto your leg. Once you have a firm grasp on the victim's wrist, use the steps in the arm extension to pull the victim to safety. The leg extension rescue can only be used to rescue an active victim because the victim must be able to grab the rescuer's leg.

Wading Assist

Do not wade into water that is deeper than your chest. Talk constantly to calm the victim. If possible, do not touch the victim directly. Extend an object (e.g., stick; pack; rifle with magazine removed, chamber empty, and muzzle pointing toward victim) that the victim can grasp. Once the victim grasps the object, pull the victim slowly to safety.

Throw

If the victim is not within reach, use an expedient line to throw a lifesaving device to the victim. A lifesaving device can be any weighted item that floats (e.g., a canteen that is one fourth full of water). The lifesaving device must be secured to the end of the rope so that the rope will feed out from its coil when tossed. Talk constantly to calm the victim. Once the victim grasps the line or the lifesaving device, pull the victim toward you at a steady pace that keeps the victim's head above the water's surface. DO NOT pull so strongly that you break the victim's grip on the line. To prepare and use an expedient line and lifesaving device—

Tie a bowline at one end of the rope. (The appendix illustrates various types of knots.)

Unfasten the lid of a canteen.

Place the bowline around the neck of the canteen.

Refasten the lid so the canteen hangs from the bowline loop.

Place one end of the rope under the ball of your forward foot to secure it, either tie a knot at the end of the rope or tie an object at the end of the rope to create a block. Stand with your weight on the end of the rope. DO NOT tie the rope around your ankle.

Coil 20 to 30 yards of the rope, and hold it in a nonthrowing hand.

Place the canteen in your throwing hand.

Use an underhand throw to pitch the canteen and rope a short distance over the victim's head. Keep your nonthrowing hand open so the coil can unfold freely. The rope should trail across the victim's outstretched arms.

Retrieve the rope if the throw is inaccurate or the victim fails to grasp it. Recoil the rope as it is retrieved.

Divide the coil and throw again.

Lifesaving Approaches

Properly approaching the victim is as important as any other aspect of a rescue. Knowing the appropriate and safest method needed to approach a victim is critical in ensuring your safety and the victim's survival. Determining the victim's physical state (distressed swimmer, victim that is active and drowning, or victim that is passive and drowning) is crucial and will determine which type of approach you will execute.

Front Surface Approach

The front surface approach is typically performed when the victim is passive. You must remember that approaching the victim from the front is dangerous because a distressed swimmer or victim that is active can lunge toward you.

Rear Approach

If the victim is active, you must remember that his response can change rapidly. Therefore, approaching him from the front could be extremely dangerous. The victim could easily grab you and take you under the water during a state of panic. Therefore, use a rear approach when possible. When approaching a victim from the rear, reassure the victim until you get to within 2 to 6 yards of the victim; at that point, stop talking to the victim in order to maintain an element of surprise.

Approach Strokes

Before approaching a victim, you must evaluate the victim:

- Is the victim a distressed swimmer?
- Is the victim active and drowning?
- Is the victim passive and drowning?

Select your approach stroke based on your evaluation. Remember, the safety of the rescuer is paramount. Do not endanger yourself in an attempt to reach a drowning victim.

Crawl Stroke Approach Stroke

The crawl stroke (free style) is the fastest approach stroke. This stroke is used when the victim is passive and/or unconscious in the water. That is, the victim is face down, submerged or near the surface of the water, is not breathing, or is not moving. If any of these conditions exist, there is an obvious need to reach the victim as quickly as possible.

Execution of the crawl stroke approach stroke is the same as the crawl stroke (see pages 1-40 through 1-44), except that your head remains above the water's surface to allow for free breathing and maintaining eye contact with the victim.

Maintaining eye contact is critical. If the victim becomes submerged, you will have a better chance of locating him if you know his last location. Your distance from the victim will determine if your head remains out of the water the entire time you are executing the stroke. If you are 54 yards or less from the victim, you should maintain eye contact by keeping your head raised until you reach the victim. If you must swim more than 54 yards to reach the victim, make eye contact with the victim, place your face back in the water for four or five strokes, then raise your head again to regain eye contact with the victim (repeat these steps until you reach the victim).

Body position
The position of your body is horizontal to diagonal and influenced by the position of the head. This allows you to maintain eye contact with the victim.

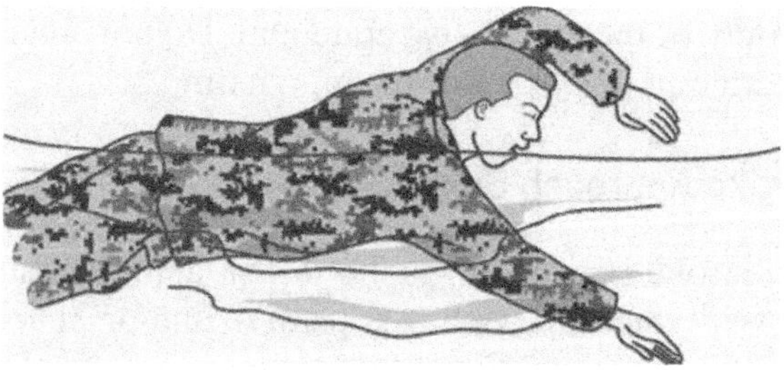

Arm action
While one arm is forward of your head to catch the water for the propulsion phase, extend your other arm alongside your body in what is known as the recovery phase. The recovery phase ends when your arm is extended forward of your head to begin the propulsion phase. In the propulsion phase, your arms are approximately shoulder-width apart and the pulling action is slightly wider and deeper to compensate for your raised head.

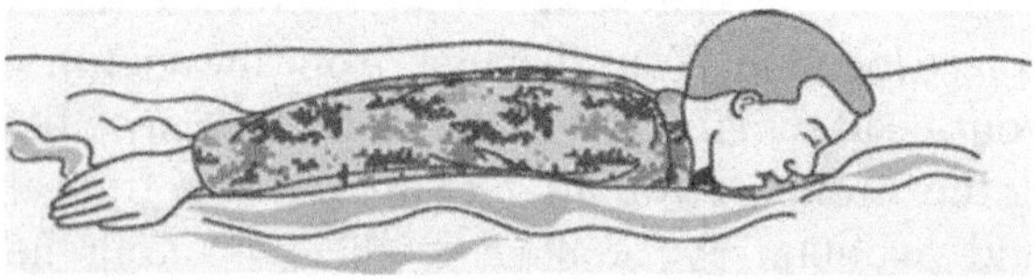

Leg action
Kick one leg in a downward motion while your other leg is recovering to the surface to prepare for the next kick (this is known as a flutter kick). This kick is used for propulsion and maintaining the lower body as horizontal to the water's surface as possible. When your head is raised, your knees are slightly bent in order to keep your feet near the surface.

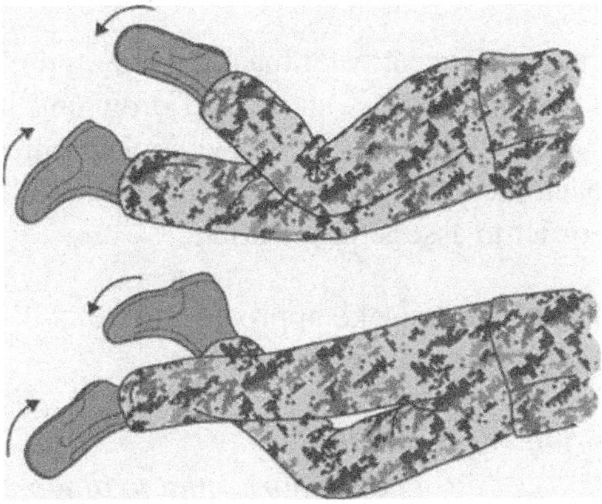

Breathing
If your face is above the water, breathe as needed (free breathing).

If traveling in excess of 54 yards to reach the victim, rotate your head to the right or left during the recovery phase and inhale when your face is out of the water or wait until your head is out of the water and facing forward while regaining sight of the victim.

Coordination
This stroke uses constant arm and leg action.

Breast Stroke Approach Stroke

The breast stroke approach stroke is used when a swimmer is distressed or a victim is active and drowning. This stroke is also used when the victim has a suspected spinal injury because it minimizes the wake created around the victim, minimizes the movement of the victim's head and neck, and helps prevent further injury.

If a victim is at a distance greater than 54 yards and if the swimmer is in distress or the victim is active and drowning in open water (ocean, lake), you should use the crawl stroke approach stroke to rapidly approach the victim, stopping approximately 11 yards from the victim in order to assess the situation.

Execution of the breast stroke approach stroke is the same as the breast stroke with a few minor deviations.

Body position
The position of your body is horizontal to diagonal and influenced by the position of the head. This allows you to maintain eye contact with the victim. It also allows you to communicate with the victim in order reassure him as you approach.

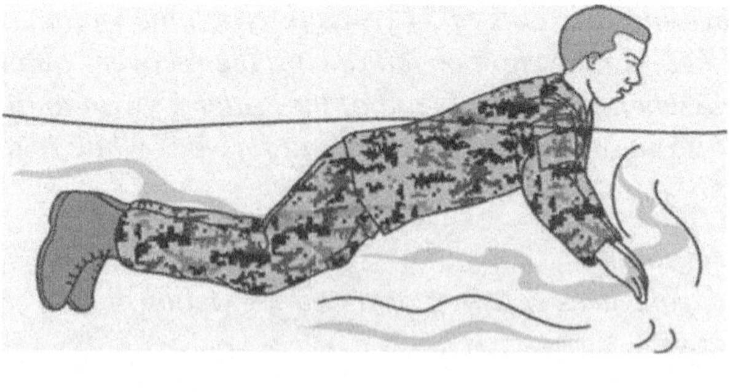

Arm action

Execute the arm action for the breast stroke, except that your arm pull is wider and deeper, which allows your head to remain above the water's surface.

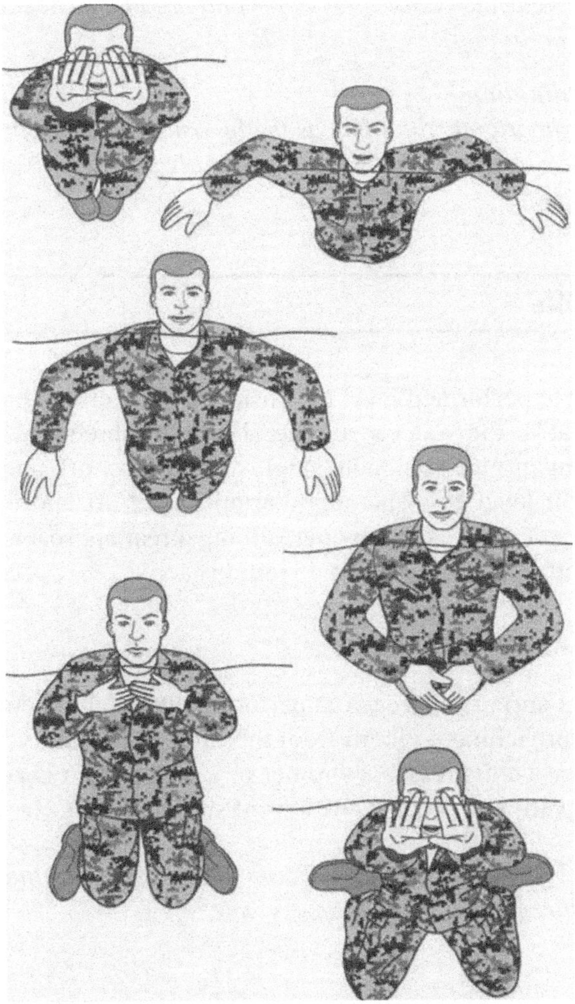

Leg action
Execute the leg action for the breast stroke, except that your legs open wider during the propulsion phase of the stroke.

Breathing
Keep your face above the water and breathe as needed (free breathing).

Coordination
Coordination is the same as the breast stroke, except that there is no glide period. You constantly stroke until you reach the victim.

Level Offs

A properly performed level off positions the victim's face up and horizontal to the water's surface. There are three types of level offs: front surface approach, single armpit level off, and the double armpit level off. The single armpit level off and the double armpit level off are done by performing a rear approach, which is the preferred approach when a victim is active.

Front Surface Approach

The front surface approach is performed when the victim is passive. Approaching a victim from the front may place you in danger because a distressed swimmer or a victim who is active may lunge toward you. To execute a front surface approach—

Stop 2 to 4 yards from the victim to reassess the situation and reassure the victim.

Determine which arm to use in order to rotate the victim into a face-up position (your right hand on the victim's right wrist or your left hand on the victim's left wrist).

Move into position and prepare to grasp the victim.

Turn sideways and move toward the front of victim.

Once in position, reach forward and grab the victim's wrist. Your thumb is on the inside of the victim's wrist, as if you were checking his pulse with your thumb. Your remaining fingers wrap around the victim's wrist.

Lean back immediately and execute a powerful scissor kick or inverted scissor kick and perform short, vigorous pulls with your free arm.

As the victim begins to move forward in the water, kick, pull, and twist outboard on the victim's wrist. The momentum created from the kick and the pulling and twisting action of the victim's wrist will rotate the victim into a face-up and horizontal position in the water.

Extend and lock out your towing arm down the length of your body and execute a wrist tow to move the victim to safety.

Single Armpit Level Off

Approach the victim.

Stop 2 to 6 yards from the victim to reassess the situation, reassure the victim, and maintain your margin of safety.

Approach the victim slowly and grab his armpit with your hand (your right hand to victim's right armpit or your left hand to victim's left armpit). Position yourself sideways to the victim and place your elbow in the center of his back.

Pull with your hand that is in the victim's armpit while pushing with your elbow that is in the victim's back to place the victim horizontal on the water's surface (face up). To assist in placing the victim horizontally, use your free arm to execute short, vigorous pulls and use your legs to execute a scissor kick or inverted scissor kick.

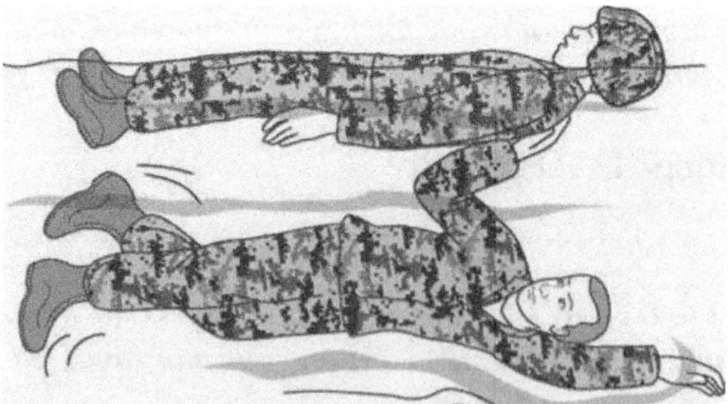

Once the victim is horizontal and on the water's surface, begin forward momentum by extending your towing arm to a fully locked-out position and executing a single armpit tow.

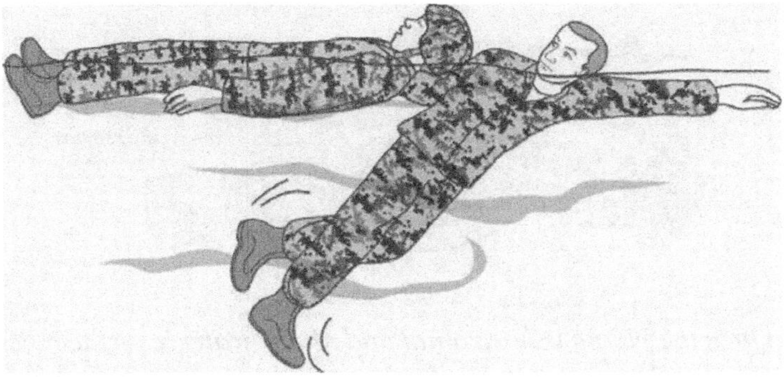

Double Armpit Level Off

Approach the victim.

Stop 2 to 6 yards from the victim to reassess the situation, reassure the victim, and maintain your margin of safety.

Approach the victim slowly and grab his armpits with both your hands. Place your elbows on his back.

Pull with your hands that are in the victim's armpits while pushing with your elbows that are in the victim's back and use an inverted breast stroke kick to place the victim horizontal on the water's surface (face up).

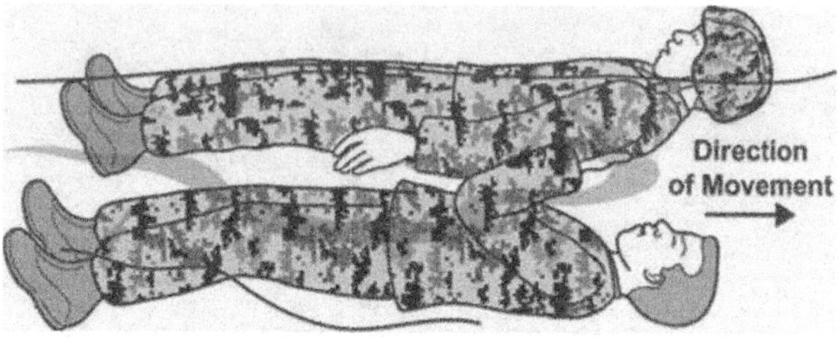

Once the victim is horizontal and on the water's surface, begin rearward momentum by slowly extending your arms into a fully locked-out position and executing a double armpit tow and inverted breast stroke kick.

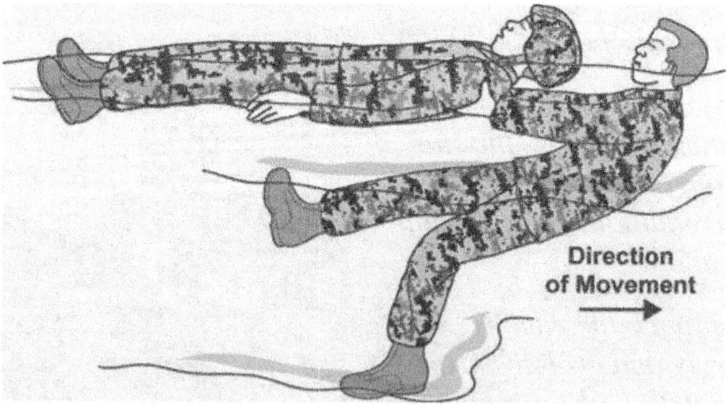

Rescue Techniques

Wrist Tow

Use the wrist tow method to rescue a victim who is floating face down. DO NOT use the wrist tow on a struggling victim. If time allows, remove your helmet and gear before attempting the rescue. Swim toward the victim using a modified breast stroke. Swim within 2 to 6 yards of the victim to maintain a margin of safety, this allows you to reassess the situation and reassure the victim. The following steps show proper front surface approach, wrist tow procedures:

Approach the victim from the front and grasp the underside of the victim's left wrist with your left hand or the right wrist with your right hand. Ensure that your thumb is on the underside of victim's wrist.

Lean back, pulling and kicking strongly to move the victim into a horizontal position.

Twist the victim's wrist to rotate the victim into a face-up position.

Swim toward safety using the lifesaving stroke.

> NOTE: The lifesaving stroke is a modified side stroke wherein the top arm is used to tow or carry a victim to safety and use a scissor kick or an inverted scissor kick is used for propulsion.

Keep a firm grip on the victim's wrist.

Keep your towing arm fully extended and along your side. This keeps the victim in the water column and prevents drag.

Ensure the victim's head does not go under water during the recovery.

Single Armpit Tow

You perform the single armpit tow after a victim has been properly leveled off. The single armpit tow uses the lifesaving stroke. To execute the single armpit tow—

Place one hand (top arm) in the victim's armpit. Your towing arm is straight and along your side.

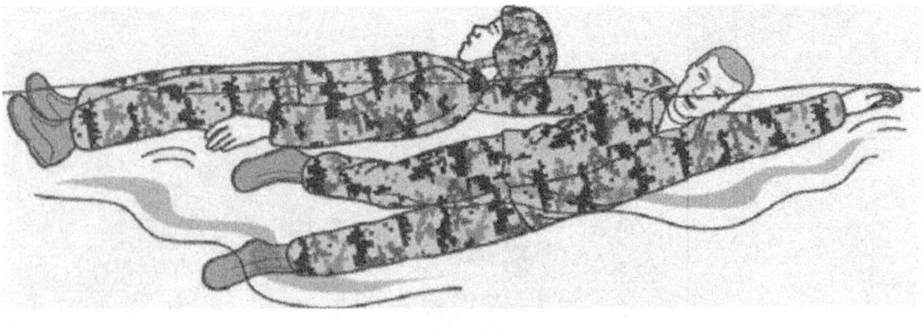

Use your lead arm (bottom arm) to execute short, vigorous pulls.

Execute a scissor or inverted scissor kick in a continuous and vigorous manner.

Use either free breathing or explosive breathing during the tow. During free breathing, keep your head above the water's surface and continuously reassure the victim. During

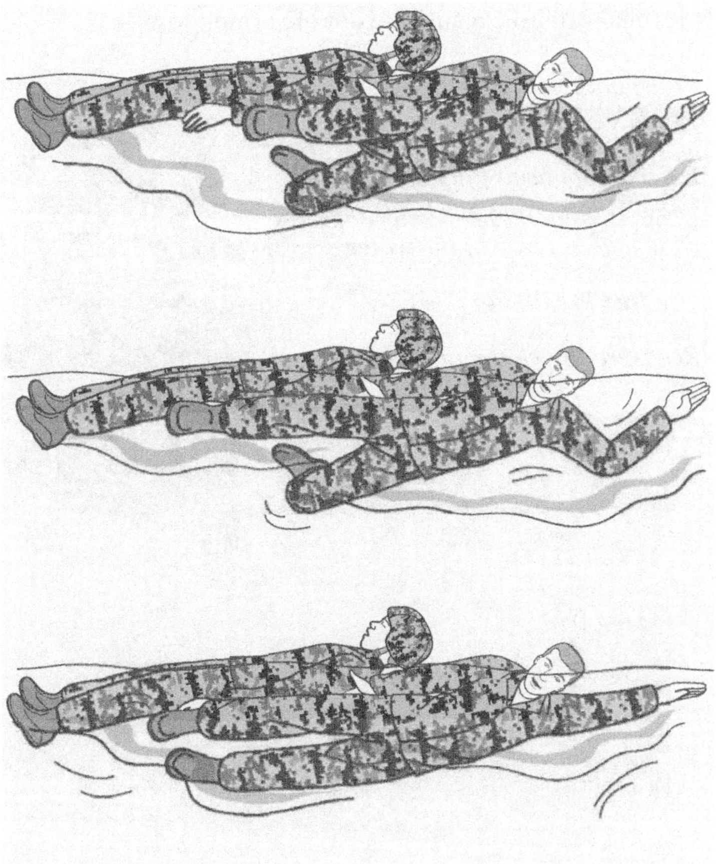

explosive breathing, put your head under the water to plane off your body's angle and reduce drag. Execute approximately two strokes, then lift your head up to breath and reassure the victim. Return your head back underwater for the next few strokes.

Double Armpit Tow

You perform the double armpit tow after a victim has been properly leveled off. To execute the double armpit tow—

Place both hands in the victim's armpits.

Extend both arms fully along your body. You are on your back, and your face is clear of the water.

Use free breathing.

Reassure the victim at all times.

Use the inverted breast stroke kick (it is the only kick that can be used due to your body position) to tow the victim to safety. (The inverted breast stroke kick is the same kick used in the elementary backstroke.) Your kick must be continuous and vigorous in order to keep the victim's face above the water.

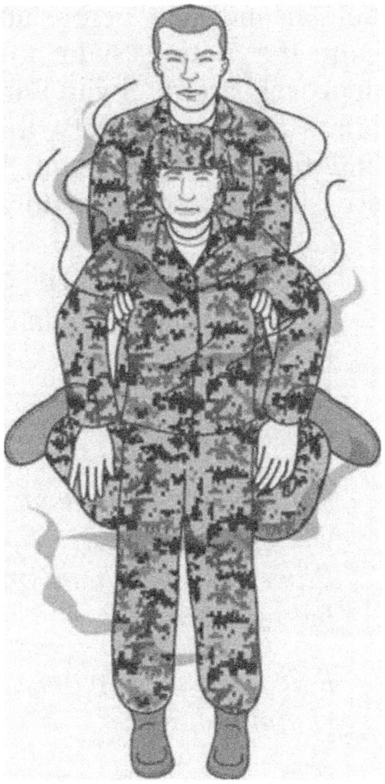

Collar Tow

You perform the collar tow after a victim has been properly leveled off. To perform a collar tow using the victim's blouse—

Maintain control of the victim by grasping his armpit with one hand and then with your free hand grasp either his combat gear or his blouse between his shoulder blades. If grasping the blouse, grasp the material with your palm up, then turn your hand over to tighten the material.

Release the victim's armpit once control is established and execute the lifesaving stroke to tow the him to safety.

Cross-Chest Carry

Use the cross-chest carry to carry a victim to safety if the victim is struggling or when moving through heavy surf. Remove your helmet and gear before attempting a rescue. Talk constantly to calm the victim. Swim toward the victim using a breast stroke approach stroke. Swim within 2 to 6 yards of the victim to maintain a margin of safety, this allows you to reassess the situation and reassure the victim. The following steps illustrate proper cross-chest carry procedures:

> **CAUTION:** The cross-chest carry causes fatigue even if you are in excellent physical condition.

Use a level-off technique to place the victim in a horizontal, face-up position.

Retain a grip on the victim with one hand. Reach over the victim with your free hand to encircle the victim's chest. Place your free hand on the victim's opposite rib cage, just below his armpit.

Release your grip once you have a secure hold on the victim's chest.

Swim toward safety using the lifesaving stroke while keeping a firm grip on the victim's chest and your hip on his back.

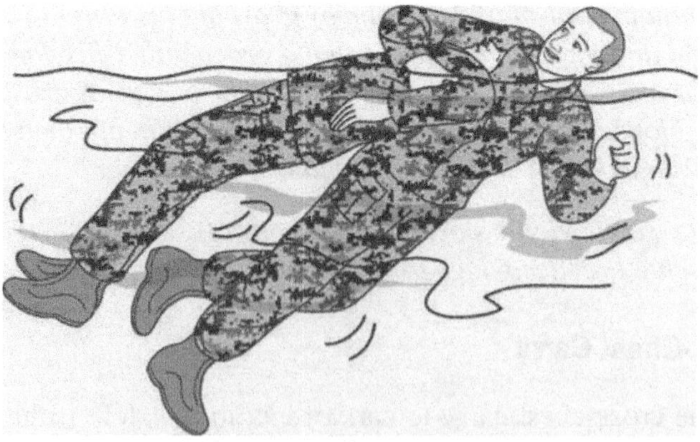

This procedure brings the victim's face and shoulders clear of the water, and typically the victim stops struggling. Sometimes, however, the victim will struggle during the swim to safety. If this happens, either tighten your grip on the victim or defend yourself with one of the techniques discussed on pages 2-25 through 2-30.

Tired Swimmer's Assist

You must maintain a 2 to 6 yards margin of safety from the victim at all times while you are getting into position to perform the tired swimmer's assist. Once behind the victim, you extend one arm, hand straight. Place your straight hand underneath the victim's opposite armpit. Maintain a 45 degree angle away from the victim, your arm is locked out. Assist the victim in propulsion until both of you reach safety.

Defense Against a Drowning Victim

DO NOT sacrifice your life in an attempt to save the victim. A struggling drowning victim poses great danger to anyone nearby. Driven by panic, the victim can grab you with great strength in an effort to climb out of the water. This can result in death for both you and the victim. The following techniques can be used to defend against a drowning victim, but the best defense against attack by the victim is to stay out of reach.

Block

The block prevents the victim from grabbing you if you have approached from the front. If the victim lunges toward you, react as follows:

> *Place one or both open hands against the victim's upper chest; being careful to avoid the victim's face, neck, and abdomen.*

Lean backwards and submerge rapidly. Keep your blocking arm(s) extended.

Swim underwater and away from the victim, and return quickly to the surface.

Stop 2 to 6 yards from the victim to reassess the situation.

Determine an appropriate course of action.

Wrist-Grip Escape/Wrist-Grip Escape Alternative

The wrist-grip escape is used when a victim grabs your arm or wrist. Submerge the victim quickly by reaching across with your free hand, pushing down on the victim's shoulder to submerge him, and kicking to propel yourself upward. While keeping the victim submerged with your hand on his shoulder, give three hard jerking pulls with your trapped hand in an attempt to break free from his grasp. Once free, swim clear of the victim and reassess the victim's condition.

If the wrist-grip escape is unsuccessful in freeing your hand, use the wrist-grip escape alternative: use your free hand to grab your trapped fist, rotate thumbs up, apply bone-on-bone contact with the victim's arm, pry your hand out of his grasp, and quickly swim away from the victim.

Front Head-Hold Escape

The front head-hold escape allows you to escape when you are facing a victim who is gripping you around your head and neck. To execute the front-head hold escape—

Take a quick breath and tuck your chin into your shoulder to protect your throat.

Clap your hands above your head (three times) to submerge instantly. This drags the victim below the water, lifts his arms from around your neck, and, typically, he releases his grasp in order to get back to the surface. If he doesn't release his grasp, apply pressure to the victim's brachial pressure points (which are located inside of the upper arm, above the elbow).

Thrust the victim's arms up and away.

Keep your chin tucked to protect your throat, and swim underwater away from the victim and return quickly to the surface at the ready position.

Stop 2 to 6 yards from the victim to reassess the situation.

Determine an appropriate course of action.

Rear Head-Hold Escape

The rear head-hold escape allows you to escape when a victim is gripping your head and neck from the rear. To execute the rear head-hold escape—

Take a quick breath. and tuck your chin down, turn your head to either side, and raise your shoulders to protect your throat.

Take a strong stroke, clap your hands above your head (three times), and submerge instantly. This drags the victim below the water and, typically, he releases his grasp in order to get back to the surface. If he doesn't release his grasp, apply pressure to the victim's brachial pressure points (which are located inside of the upper arm, above the elbow).

Thrust the victim's arms up and away.

Twist your head and shoulders until free.

Swim underwater away from the victim and return quickly to the surface.

Stop 2 to 6 yards from the victim to assess the situation.

Determine an appropriate course of action.

Administering First Aid/Rescue Breathing

To administer first aid/rescue breathing, remove the unconscious victim from the water when possible and stop any bleeding.

First, open the airway and check for breathing. If the victim is not breathing, give two breaths of air. Check for a pulse. If there is a pulse but the victim is not breathing, continue rescue breathing. If there is no pulse, start cardiopulmonary resuscitation (CPR). For detailed information on administering CPR and rescue breathing, refer to American Red Cross CPR skill cards.

WARNING

> If the victim has no pulse and is not breathing, administer CPR immediately. If the victim does have a pulse but is not breathing, give rescue breathing only. If the victim has a pulse and is breathing, DO NOT give CPR—CPR could prove fatal.

Next, protect any wounds from exposure, and treat for shock.

Chapter 3
Treatment of Casualties and Avoidance of Dangerous Marine Life

To survive in the water, you will face many challenges other than just being able to swim. For example, injuries and/or fatigue can lead to drowning and exposure to the elements can lead to hypothermia or heat injuries and may require medical attention. Finally, your presence in the water may be seen as a threat to or food source for a variety of marine life and you may be attacked. This chapter advises you on how to minimize these risks in order to stay alive.

Drowning

Drowning is suffocation by a liquid. A drowning victim inhales water into the lungs or the throat closes by reflex so that little or no water can enter the windpipe. In either case, a victim can no longer breathe.

Symptoms

One symptom of drowning is that the victim may call for help and has an expression of dread or panic. But typically a victim that is active and drowning may not call for help because he is trying to conserve his air and will not speak.

Another symptom of drowning is when the victim thrashes at the water's surface. If the victim stops or grows calmer, he has likely been overcome by fatigue, hypothermia, or a lack of air. At this stage, the victim usually has 20 to 60 seconds before going under the water's surface.

Treatment

If the victim is not breathing, begin rescue breathing. Place the victim on his back, tilt head back to open airway, pinch the nose, and give two full breaths. If the victim does not inhale during the first two breaths, reposition his head and attempt two more breaths. Check for a pulse. If a pulse is present, but the victim is still not breathing, continue rescue breathing. If a pulse is not present, begin CPR. See MCRP 3-02G, *First Aid*, for rescue breathing and CPR details.

WARNING

If the victim has no pulse and is not breathing, administer CPR immediately. If the victim does have a pulse but is not breathing, give rescue breathing only. If the victim has a pulse and is breathing, DO NOT give CPR—CPR could prove fatal. Continue first aid until medical help arrives.

A victim who is not breathing and has no pulse may appear dead. DO NOT decide that death has occurred. Continue with the prescribed treatment. A corpsman or medical officer should decide whether the victim can or cannot be revived.

Hypothermia

Hypothermia is the abnormal lowering of the body's internal (or core) temperature to 95 degrees or below. It occurs when the body loses heat faster than it produces heat. A hypothermia victim loses the ability to move quickly, becomes mentally sluggish, slips into semiconsciousness, lapses into a coma, and dies when internal body temperatures drop too low.

The chilling effects of cold air, wind, or water can produce hypothermia. Water poses the greatest threat because it transfers heat 25 times faster than air. Depending on the water's temperature, a victim can succumb to hypothermia within a few minutes. The body's sudden contact with cold water can also set off a body reaction known as the mammalian diving reflex. This reflex can greatly increase survival time (especially for women and children) in or under cold water. The mammalian diving reflex shuts off blood circulation, except for the flow between the heart, lungs, and brain. The small amount of oxygen left in the blood and lungs is saved for the body's vital organs. This reflex has allowed people to survive being under cold water for an extended period of time. Therefore, a cold water drowning victim should be treated as if still alive even though the victim is not breathing, has no pulse, and may appear dead. DO NOT decide that death has occurred. Continue with the prescribed treatment. Victims of hypothermia can appear to be dead when they are not. A corpsman or medical officer should decide whether the victim can or cannot be revived.

WARNING

If the victim has no pulse and is not breathing, administer CPR immediately. If the victim does have a pulse but is not breathing, give rescue breathing only. If the victim has a pulse and is breathing, DO NOT give CPR—CPR could prove fatal. Continue first aid until medical help arrives. Check for a pulse for at least 45 seconds.

Symptoms

Once the body's core temperature drops, the victim will show one or more of the following symptoms:

- Violent and uncontrollable shivering as the body tries to warm itself.
- Slow or slurred speech.
- Disorientation or poor coordination.
- Loss of skin color.
- Blue and pinched lips.
- A slowing or stopping of shivering that progresses into a rigid torso and limbs.

Survival Time

A hypothermia victim's survival depends on the water's temperature and the time spent in the water. A small body build cools faster than a large build. Children cool faster than adults.

To increase your chance of survival in the water, utilize the HELP position described on page 1-37. Extra clothing and inactivity (remaining motionless in the water) can also increase your survival time.

Treatment

A hypothermia victim must be warmed to prevent further heat loss; therefore, treatment should begin as soon as possible. Consciousness of the victim determines the treatment that should be pursued. See MCRP 3-02G for specific treatment information. The following treatment procedures are recommended:

- If the victim is conscious, give the victim warm fluids. Give candy or sweetened foods to a victim who is able to eat.
- If the victim is unconscious, place him on his back with his head tilted back to ensure an open airway.
- DO NOT massage the victim. Massage can break blood vessels and create swelling, internal pressure, and blocked blood circulation.
- DO NOT give alcohol to the victim. Alcohol lowers the victim's body temperature.
- Shock is a possibility, treat accordingly.
- Seek medical help immediately.

If you are able to remove the victim from the water, apply the following steps when possible:

- Get the victim into shelter.
- Remove the victim's wet clothing.
- Put the victim in dry clothing.
- Place the victim in a sleeping bag if one is available. It may be necessary to place another Marine in the sleeping bag with the victim.
- Place as much insulation as possible between the victim and the ground.
- Use hot water bottles, electric blankets, or blankets heated in an oven or by a campfire to warm the victim's neck, groin, and the sides of the chest.

CAUTION: DO NOT apply heat to extremities.

Heat-Related Injuries

Heat-related injuries include heat cramps, heat exhaustion, and/or heat stroke. They can occur when a Marine—

- Is exposed to extreme heat, such as from the sun or a combination of high air temperatures and water temperatures.
- Does not wear proper clothing or gets overexposed to the sun.
- Becomes dehydrated.

With heat cramps, muscles may cramp in the arms, legs, and/or stomach. The victim may sweat excessively. To treat, create improvised shade for the victim and have him drink water.

With heat exhaustion, the victim sweats heavily; presents pale, moist, cool skin; and complains of a headache, weakness, dizziness, and/or loss of appetite. A victim may also experience heat cramps, nausea, vomiting, an urge to defecate, chills, rapid breathing, confusion, and a tingling sensation in the hands or feet. To treat, pour water over the victim and fan him to speed up the coolant effect of evaporation, have the victim drink freshwater, and attempt to provide shade.

With heat stroke, the victim stops sweating, which results in red, flushed, hot, dry skin. The victim may first experience headaches, dizziness, nausea, fast pulse and respiration, and/or seizures and mental confusion. The victim can become unconscious and die if not treated quickly. To treat, create improvised shade for the victim. Have him drink water and elevate both legs if possible.

Burns

Apply a field dressing (or the cleanest material available) to the burn. Give sips of water to a casualty who is conscious and not nauseated. When treating burn casualties—

- DO NOT remove clothing stuck to the burns.
- DO NOT break any blisters.
- DO NOT apply grease or ointments to the burns.

For electrical burns, check for both an entry and exit burn from the passage of electricity through the body. An exit burn may appear on any area of the body, not necessarily opposite the entry burn.

For burns caused by wet or dry chemicals, flush the burns with large amounts of water and cover with a dry dressing.

For burns caused by white phosphorus, flush the area with water, then cover with a wet material, dressing, or mud to exclude air and keep the white phosphorus particles from burning.

For laser burns, apply a field dressing.

Common Medical Problems Associated with Sea Survival

Seasickness

Seasickness is the nausea and vomiting caused by the bobbing motion created by the wave action of a flotation device. Seasickness can result in—

- Dehydration and exhaustion.
- A loss of the will to survive.
- Others becoming seasick.
- Unclean conditions.

To treat seasickness—

- Wash both the Marine and the flotation device to remove the sight and odor of vomit.
- Keep the Marine from eating food until the nausea is gone.
- Have the Marine lie down and rest.
- Give the Marine seasickness pills if available. If the Marine is unable to take them orally, the pills should be inserted rectally for absorption by the body.

Saltwater Sores

Saltwater sores occur when skin that has abrasions or is cut is exposed to saltwater. The sores may form scabs and pus. Do not open or drain. Flush the sores with freshwater, if available, and allow to dry. Apply antiseptic, if available.

Blindness/Headache

Irritants or the effects of the sun's rays reflecting off the water can cause temporary blindness or headaches. If flames, smoke, or other irritants get in your eyes, flush the eyes immediately with saltwater, then with freshwater, if available. Apply an ointment, if available. Bandage both eyes for 18 to 24 hours or longer if the damage is severe. If glare from the sky and water causes your eyes to become bloodshot and inflamed, bandage the eyes lightly. Try to prevent this problem by wearing sunglasses or goggles with a sunglass insert.

Constipation

This condition is a common problem associated with dehydration. For constipation, do not take a laxative if it is available; this causes further dehydration. Drink freshwater, if available.

Sunburn and Dehydration

The sun's rays reflect at all angles off the waves of the water; therefore, sunburn and dehydration are serious problems in sea survival. Try to prevent sunburn by—

- Erecting an improvised canopy, with available floating materials, to provide shade.
- Wearing your soft cover or using a cloth, such as a handkerchief, to cover your head.
- Covering your skin with sunscreen or lip balm from your first aid kit. Your lips, nostrils, eyelids, the backs of your ears, and the skin under your chin sunburns easily. If enough sunscreen cream is available, all exposed skin should be covered.

Dehydration is caused by the loss of the body's vital fluids. Dehydration in saltwater may result from a combination of factors such as a lack of water, the effects of saltwater on skin tissue, sunburn or vomiting from seasickness, and other causes. Sleep and rest and reduced water and food intake are the best ways of enduring periods of exposure. The following measures will delay the effects of dehydration:

- DO NOT drink saltwater.
- DO NOT drink urine.
- DO NOT drink alcohol.
- DO NOT smoke.
- DO NOT EAT unless water is available.

Dangerous Marine Life

You may see many types of marine life around you, and some are more dangerous than others. Generally, sharks are your greatest danger, followed by barracudas. However, most marine life will not deliberately attack a human. The most common injuries from marine life are wounds from bites, stings, or punctures. With the exception of sharks and barracudas, most injuries are a result of either trying to catch game or from contact abrasion with marine life. To treat an injury that results from a dangerous marine animal bite—

- Pack the wound with gauze and then apply a pressure bandage, if available.
- Treat for shock.
- Obtain medical attention as soon as possible.

For a more detailed information on the treatment of casualties and dangerous marine life, see MCRP 3-02F and MCRP 3-02G.

Sharks and Barracudas

Only about 20 percent of all shark species are known to attack people. Sharks have an acute sense of smell, and the smell of blood in the water will draw them to their prey. They are also very sensitive to any abnormal vibrations in the water; therefore, the sound caused by a struggling swimmer or underwater explosions will attract them.

Barracudas are bold and inquisitive fish. They have been known to attack men who are wearing shiny objects. Barracudas may charge at lights or shiny objects at night.

A group of swimmers can maintain a 360 degree watch while in the water. Therefore, to protect yourself from sharks and barracudas, stay with other swimmers. The group members can either frighten away or fight off sharks or barracudas better than one person can. Keep all your clothing on, including footwear. Historically, sharks attack unclothed individuals in groups first, mainly in the feet. Clothing also protects you against abrasions from a shark's tough skin should the shark brush up against you. Avoid urinating heavily, let urine dissipate between discharges. If you must defecate, do so in small amounts and toss it as far away from the group as possible. Do the same if you must vomit.

If attacked, the use of firearms by swim sentries should be used with extreme caution because of the risk of injury to other swimmers. If unarmed or unable to make an improvised weapon, kick and strike the shark. Avoid using your bare hands to strike the shark, injury can result to your hands due to a shark's tough skin. Target areas on a shark are the gills, eyes, and underbelly. Blows to the snout are also not recommended because a shark will tilt its head up and thrust its jaws forward when biting.

Sea Snakes

Sea snakes are venomous and sometimes found in mid-ocean. They are unlikely to bite unless provoked. Your best protective measure is to avoid them.

Poisonous Fish

Many reef fish are toxic, and can kill you if eaten. Generally, poisons are present in all parts of the fish, but especially concentrated in the liver, intestines, and eggs.

Turtles and Moray Eels

Turtles and moray eels normally inflict minor bite wounds. Treat this type of injury by cleaning the wound.

Corals

Coral, dead or alive, can inflict painful cuts. Clean all coral cuts thoroughly. DO NOT use iodine to disinfect any coral cuts. Some coral polyps feed on iodine and may grow inside the flesh.

Jellyfish, Portuguese Man-of-War, Anemones, and Others

This group of marine animals inflicts injury by stinging their victims with their tentacles. Contact with their tentacles produces burning pain, a rash, and small hemorrhages on the skin. Shock, muscular cramping, nausea, vomiting, and respiratory distress may also occur. Gently remove the clinging tentacles with a towel and wash or treat the area. Use diluted ammonia or alcohol and talcum powder to treat the injury if available.

Spiny Fish, Urchins, Stingrays, and Cone Shells

These animals inject their venom by puncturing the skin with their spines. General signs and symptoms include swelling, nausea, vomiting, generalized cramps, diarrhea, muscular paralysis, and shock. Deaths are rare. Treatment consists of soaking wounds in hot water, if available, to deactivate heat-sensitive toxins.

Chapter 4
Negotiating Water Obstacles

Marines face water obstacles in saltwater, freshwater, and brackish water (where freshwater and saltwater meet). These water environments differ considerably and pose distinct problems for Marine tactical units and swimmers. Saltwater obstacles include tides, surfs, and currents. Brackish water obstacles include back bays. Freshwater obstacles include rivers and canals.

Tides

Tides are periodic changes in the surface levels of oceans, bays, gulfs, inlets, and rivers. The Moon's and Sun's gravitational pulls cause tides. Depending on the situation, tides can either help or hinder Marines in their endeavors to conduct amphibious or riverine operations. Tides can provide sufficient depths to allow for the passage of landing craft or boats over reefs, trees, and other underwater obstructions. On the other hand, tides can render a river fording site unusable to tactical vehicles. Direction, level of change, and amount of change determine tidal nomenclature.

Tides that show change in direction are flood tides and ebb tides. Rising tides are known as flood tides. Falling tides are known as ebb tides.

Tides that show extreme levels of change are high tides and low tides. High tide is the period when water is at its greatest depth. Low tide is the period when water is at its most shallow depth.

Tides that show the least and most amount of change are neap tides and spring tides, respectively. Neap tides have the least amount of change in water levels between high and low tide. Neap tides occur at the half moon, when the Sun and the Moon are aligned at a 90-degree angle with the Earth. In this position, the Sun's and Moon's gravitational pulls offset each other. Spring tides have the highest floods and lowest ebbs. Spring tides occur at or shortly after the new moon or full moon when the Sun, Moon, and Earth are approximately in line. In this position, the Sun's and Moon's gravitational pulls are combined.

Surf

Waves break upon entering shallow water and create surf. The offshore area where waves break is the surf zone, which can present many hazards. Breaking waves often trap air bubbles and create a foamy appearance. Bubbles lower the water's density and decrease buoyancy. Move through foamy surf as quickly as possible. The type of wave determines your survival technique.

Wave action moves you toward shore. Lie on your back or side with your head pointing in the direction of the beach and your feet pointing into the waves.

As one wave approaches the beach, another drains away; relax and do not swim against the draining water. When a new wave is within about 3 yards, start swimming toward shore. Continue to swim until the wave lifts you and moves you toward the beach. Once the wave loses forward momentum, relax and repeat the cycle. If nearing rocks, turn your body and approach feet first to reduce the chance of striking your head and arms.

Plunging Waves

A plunging wave is a breaker whose crest curls forward and falls ahead of its base. Because of its power and underwater turbulence, a plunging wave poses the greatest surf threat. If caught in a plunging wave, you can be pulled underwater and pitched about violently. This can cause you to panic, which can increase your chance of drowning. Perform the following steps to escape a plunging wave:

Tuck into a ball with your head against your knees and your forearms locked around your legs, just below the knees.

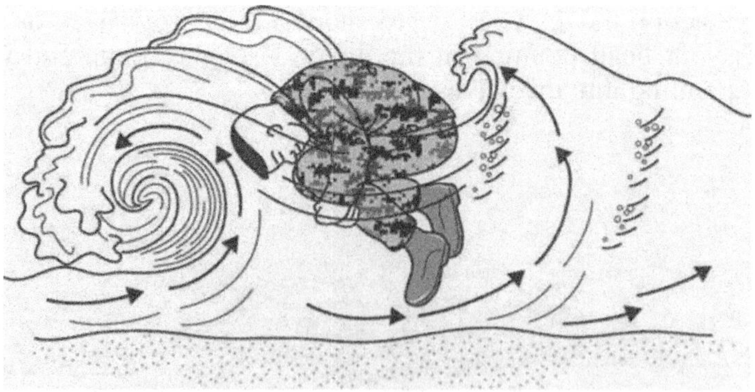

Relax in this position until the turbulence subsides and you float to the surface. This can take 30 seconds or more.

Swim toward shore.

Note: If threatened by another plunging wave, dive underwater into the wave.

Spilling Waves

A spilling wave does not break. Instead, its crest slides forward without curling. A spilling wave creates less turbulence and poses less of a threat than a plunging wave. If caught in a spilling wave, relax and let the wave carry you to shore.

Surging Waves

A surging wave occurs on a beach with a steep underwater gradient. It never really breaks, but the crest rises while the base slides up the beach with great force and speed. Once the wave reaches its highest point on the beach, it rushes back as quickly as it surged forward. If you are standing on the bottom when a surging wave advances or retreats, the wave can knock you off your feet and pull you into the surf zone. Do not try to stand or walk on the bottom. Swim toward the beach as soon as possible.

Currents

Offshore Currents

An offshore current occurs outside the surf zone. Typically, it occurs at bay entrances, in island channels, and between islands and the mainland. An offshore current flows parallel to or away from shore. If the offshore current is created by tides, its current strength and direction vary at different times of the day.

If caught in an offshore current, you may be carried in a direction you do not want to go. DO NOT try to swim directly to safety. If the current is moving directly away from the shore, relax and wait until the current dies out or turns toward land. Once the current subsides, use any survival stroke to swim toward shore. If the current is moving parallel to shore, use any stroke to move at an angle across the current and toward shore.

Rip Currents

A rip current occurs when waves pile water against the shore faster than the water can drain. The water flows rapidly along the beach until it is deflected seaward by a bottom obstruction. Then the rip current flows through the surf zone and into open water at a speed of up to 2 knots. This action can cut deep trenches in the sand. A rip current dies out once it reaches open water (usually within a few hundred yards of the shore).

A rip current can pull you out to the open sea. If caught in a rip current, DO NOT try to swim against the current. A rip current moves faster than most people swim, and it is impossible to swim to shore once caught in it. Relax and stay afloat until the current runs out. Once the current subsides, use a survival stroke to move parallel to the shore until you are out of the current, then begin swimming toward shore.

Littoral Currents

A littoral current occurs when a wave breaks against a beach at an angle. This current flows parallel to the shoreline and does not pose a great threat. If caught in a littoral current, use the combat travel stroke to swim across it at an oblique angle.

Back Bays

Once on the beach, you often face one or more rows of low hills called dunes. Behind the dunes, you may encounter a low-lying stretch of ground thickly covered with scrub trees and bushes. This area gives way to wetlands known as back bays. Back bays consist of muddy islands that are almost submerged during flood tides and separated by channels of brackish water of varying depths. Channel bottoms usually contain soft mud. Back bays pose major obstacles to vehicular traffic.

Infantry can cross back bays, but only with great effort. If crossing back bays by foot, consult detailed navigation charts and use the following guidelines to plan your route:

- Avoid water less than waist deep; walking in shallow water or soft mud is extremely tiring.
- Avoid back bay islands; these low-lying islands are usually too muddy to support foot traffic.
- Seek out deep water; floating with a pack is less tiring than walking through shallow water or soft mud.
- Seek out sand, shell, gravel, or stone bottoms; these firmer bottoms generally ease travel and help conserve energy.

Rivers and Canals

A river is a large, natural stream of water that empties into a larger body of water. The slope of the riverbed and the volume of water in the river determine its current.

Canals resemble small rivers or streams in their width and depth, but usually lack any significant current. Climbing out of these waterways can be difficult if the canal is flanked by steep banks.

Chapter 5
Fording Waterways

WARNING

Fords are dangerous. Cross as quickly as possible.

A ford is any site in a river, stream, or canal where the water is shallow enough for troops or vehicles to cross without using flotation devices. Canal bottoms are usually too soft to support fording vehicles, and wading infantry frequently stumble.

The tactical situation dictates the location of the fording site. Seek fords that are protected from enemy observation and that allow for adequate supporting fires. A night fording takes at least one and a half times as long as a daylight fording.

> *NOTE:* In this chapter, the term "river" refers to rivers, streams, and canals.

Selection of a Ford Site

The following table identifies desirable fording characteristics:

Characteristics	Comments
Concealment	The ford hides personnel and vehicle movement from enemy observation.
Accessibility	The ford should have low banks with gentle gradients. This allows a free flow of traffic at both the entrance and the exit.

Characteristics (Continued)	Comments (Continued)
Slow Current	The ford's current should not exceed 1.5 meters per second if possible.
Firm Footing	The ford's bottom, entry, and exit composition should be firm enough to support traffic. Do not drive a vehicle over any bottom composition that a 2-inch diameter stick can be pressed into more than 1 or 2 inches.
Gently Sloped Channels	The ford's entry and exit points should be gently sloped. If possible, locate a portion of the stream where the channel is not actively shifting.
Depth	The fording depth is less than or equal to the least capable vehicle.

Determine the Slope

Units move into and out of water faster and more quietly if entry and exit points are not steep or muddy. Slope is the amount of change in ground horizontal distance (run) and in vertical elevation (rise) from one point to another. Slope is usually expressed as a percentage. You can use a clinometer, map, or line of sight and pace to measure the percentage of a slope.

Clinometer

A clinometer measures percentage of a slope. It is a component of the M9 armored combat excavator, and organic to many Marine engineer units. However, engineers are also taught (and often use) field-expedient methods to determine slopes and gradients.

Map

A map measures the horizontal distance along a desired path. Determine the difference in elevation between the path's starting and ending points. From a map's scale, you can determine the distance between two points. From a map's elevation lines, you can determine the difference in height between the same two points. Both figures must be in the same unit of measure (e.g., feet, meter). Divide the elevation (rise) by the distance (run) and multiply by 100.

Rise = 165 - 120 = 45 meters

Run = 200 meters

% Slope = $\frac{45}{200}$ x 100 - 22.5%

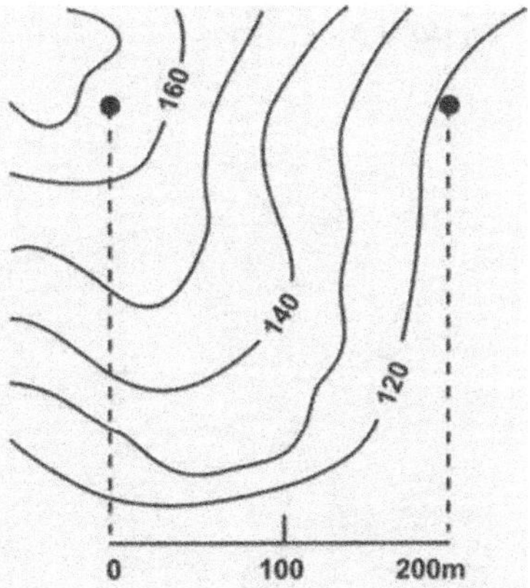

Line of Sight and Pace

To determine line of site and pace, stand at the bottom of the slope, keep your eyes level, pick a spot on the slope, then pace the distance. The number of paces multiplied by a standard measure of 0.75 meter determines the run. The eye-level height (usually 1.5 to 1.7 meters) determines the rise. Repeat this procedure until you have covered the entire distance you want to measure for each spot (vertical and horizontal). Add the vertical distances to provide total rise and the horizontal distances to provide total run.

Determine the Current Speed

Current speed increases as channels narrow. It may be necessary to locate a wider ford location to obtain a slower stream current. The following steps are used to calculate the speed of the current:

Determine points A and B along a channel. Then measure the distance between those two points.

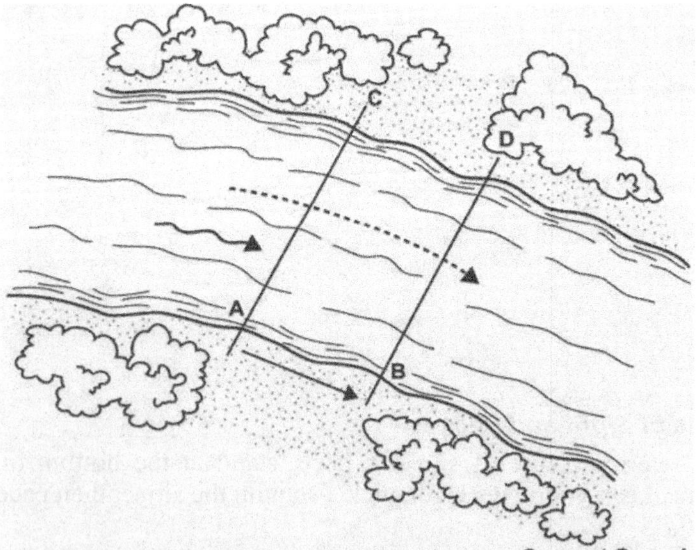

$$\frac{\text{Distance (meters) Between Points A and B}}{\text{Time (seconds) Between Points C and D}} = \text{Current Speed (meters per second)}$$

Sight directly across the water from points A and B to locate points C and D.

Throw a floating object (e.g., a stick) upstream from points A and C. Observe the object as it floats toward points B and D.

Subtract to find the time it takes for the object to float from start to finish.

Do not attempt to swim across currents that are moving faster than 1.5 meters per second. Equivalents of this speed include—

- Quick-time march rate of 120 counts per minute with one, 30-inch step at each count.
- 5 feet per second.
- 3.5 miles per hour.
- 5.5 kilometers per hour.

Measure River Width

A river's width can be estimated from the width of its symbol on a scaled topographic map. If this is not possible, use the following compass techniques:

Stand at the water line (A).

Shoot an azimuth to a point on the opposite bank (B).

Move upstream or downstream until you are at a point (C) where you can shoot an azimuth 45 degrees larger or smaller than the original azimuth.

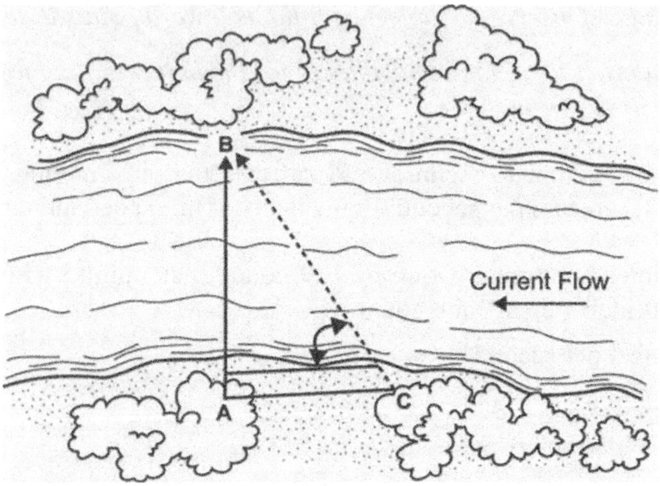

Measure the distance between points A and C. The distance calculated equals the river's width.

Calculate Downstream Drift

A river's current causes personnel and equipment to drift downstream. If personnel and equipment are aimed straight across the river, they will sideslip downstream as they move across the current to the other shore. Therefore, personnel and equipment crossings must compensate for the effects of a river's current and entries are usually made upstream of the desired exit point. Use the formula below to calculate downstream drift.

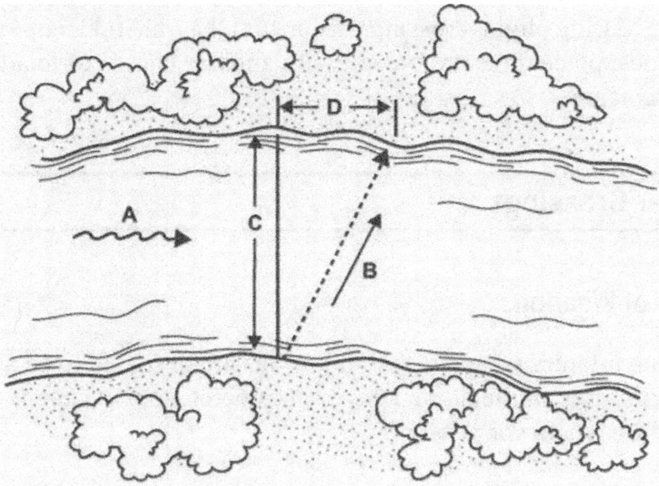

$$\frac{\text{Current Speed (A)}}{\text{Crossing Speed (B)}} \times \text{River Width (C)} = \text{Downstream Drift (D)}$$

> *Note:* The crossing speed for a swimmer across a river may vary but is generally limited to 1 meter per second. All measurements must be in the same unit of measure (e.g., meters, feet).

The Buddy System

Whenever a Marine unit must enter into or operate on the water, a "buddy system" is employed in which every Marine is paired with a swimming partner. The buddy system matches an experienced swimmer with a weak swimmer. The experienced swimmer assists and encourages the weaker swimmer and bolsters confidence during night crossings. If a unit has an odd number of Marines, place the extra person with another pair to form a three-person team.

Water Crossings

Care of Weapons

Marine infantry weapons and munitions are designed to be able to operate after immersion. However, protect your weapons from moisture whenever possible.

A gas-operated weapon can malfunction if water travels down the barrel and enters the gas tube. To protect the gas tube—

Close the weapon's bolt before entering the water.

Seal the muzzle with a condom, balloon, plastic spoon wrapper, or other form of waterproof material.

Tie or melt the protective cover to create a watertight seal.

When the muzzle's protective cover is no longer needed, remove it. Open the bolt and inspect the barrel. If the tactical situation permits, swab excess moisture from the barrel. Test fire automatic weapons, if possible. Field strip and clean weapons as soon as possible. If time does not allow for a complete inspection, rinse inaccessible areas with small amounts of diesel fuel, then dry.

ISOMAT Raft

Construction of an ISOMAT raft is time-consuming. This type of raft should not be employed as part of an attack, but used for logistical purposes (e.g., evacuating stretcher cases, transporting supplies). Use the following steps to build an ISOMAT raft:

Wrap ISOMAT sleeping pads around sturdy sticks.

Use parachute cord and square knots to tie the pads securely in place and to lash stick ends together in a rectangle.

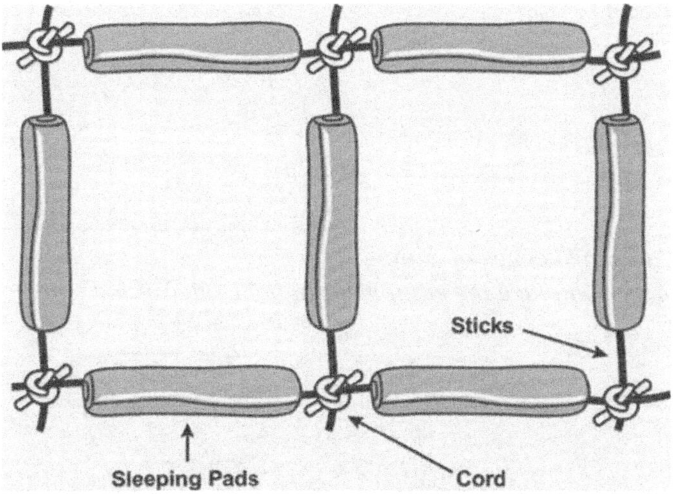

Note: The ISOMAT raft pictured can support several hundred pounds. However, the cargo will get wet if not properly waterproofed.

Poncho Raft

A poncho raft can support two Marines and their equipment and is well suited for long crossings. Use the following steps to build a poncho raft:

Inspect two ponchos and ensure they are serviceable.

Lay one poncho flat on the ground, with the hood-side up.

Cinch the hood tightly to form a gooseneck or tie in a knot.

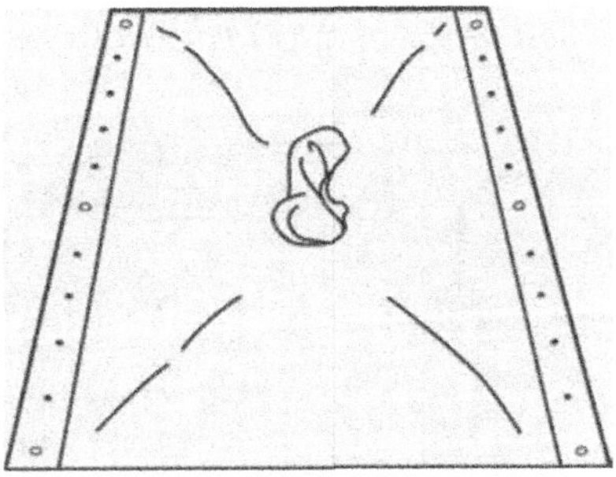

Pad sharp edges of equipment and place the equipment in the center of the poncho.

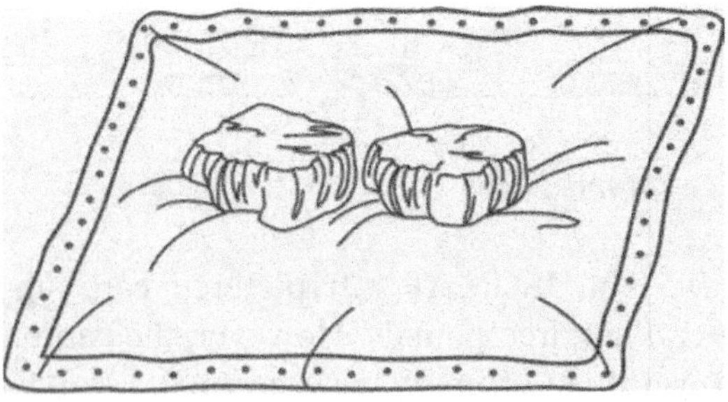

Place the second poncho over the equipment, rubber side up, and hood facing down.

Snap the edges of the two ponchos together.

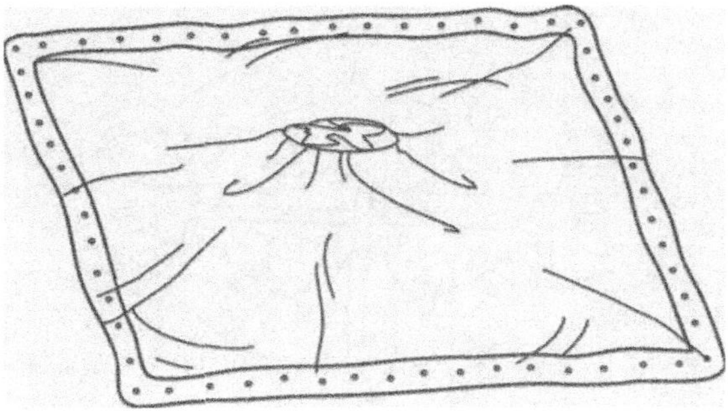

Roll the edges toward the equipment.

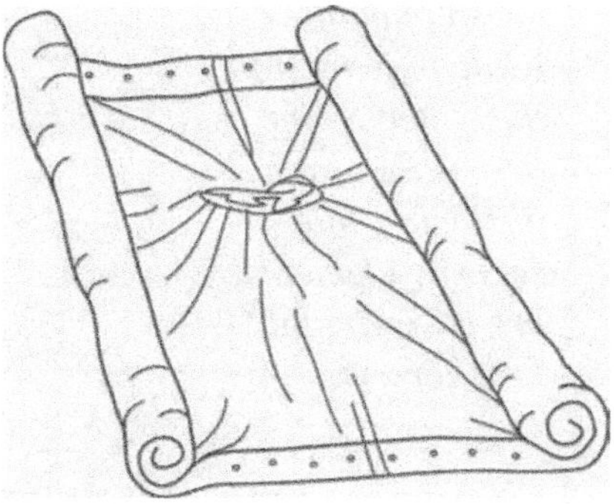

Roll the edges into pigtails and tie them off.

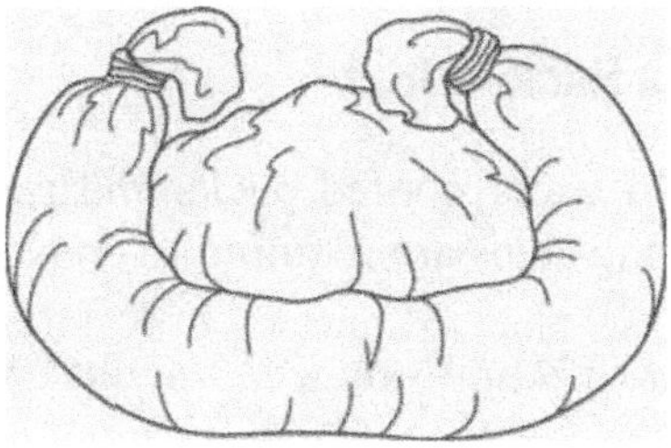

Pull the pigtails together over the top and lash them securely.

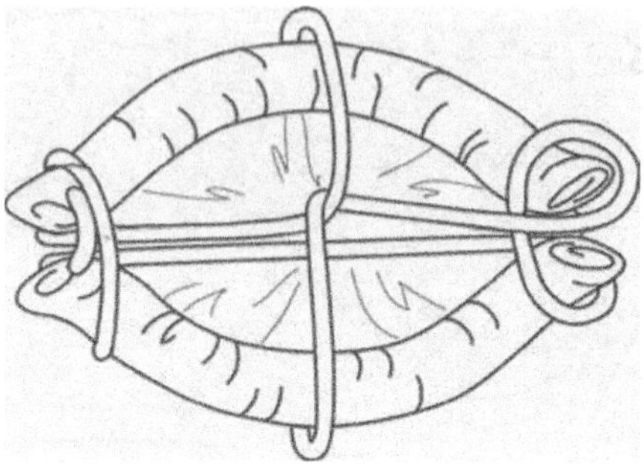

Protect the raft from brush punctures while placing it in the water. Swim across the water obstacle while security elements are covering the far shore.

Construction of a Pack Raft

You will need two waterproofed packs and two M16A2 service rifles. The following steps are required to construct a pack raft:

Place two packs side-by-side with the pack frames on the deck. The tops of the packs are opposite of each other.

Loosen the main compartment straps on both packs.

Insert one rifle on each end between the straps and the packs. The muzzles are opposite of each other. The rifles serve as one means to secure the packs together. Place the front sight post under the top flap.

Tighten the straps so that the rifles and packs are secure.

Take the excess strap on the inner side of each pack and secure it to the opposite pack to better secure the two packs together.

Take the excess straps on the outer sides of the packs and use those straps as safety lashing for the rifles.

Tuck the excess straps and check to make sure the rifles and packs are secure.

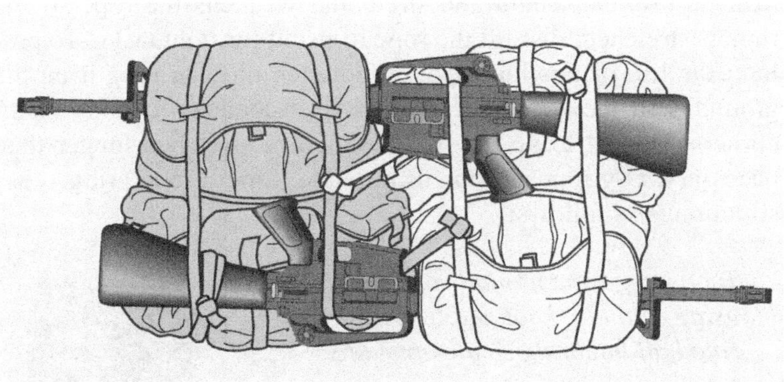

Single-Rope Bridge

A single-rope bridge offers a temporary and quick way to cross small rivers. It also provides extra security while crossing swift waters. At night, it prevents straggling, and guides units precisely from one side of the river to the other side. If crossing a river at night, plan for at least one single-rope bridge.

If your unit is crossing a river with swift currents or water depths above 4 feet, the unit is carrying sufficient rope to span the crossing site, and the tactical situation permits, secure the rope on near and far banks to provide a hand-hold for crossing Marines. This reduces the time required for the entire unit to cross and provides a degree of comfort/confidence for poor swimmers. Use a squad-sized bridge team to construct a single-rope bridge. Station several strong swimmers at the water's edge to help anyone who has trouble crossing.

Nylon rope is normally coiled in 120 foot lengths. It is 0.6 inches in diameter and has a breaking strength of about 3,840 pounds. Over time, a nylon rope can stretch to as much as one-third more than its original length and stretching weakens the rope. If the rope is stretched, discard the rope or use it for light tasks. To prolong the life of a nylon rope, do not step on it or drag it on the ground. Pad the rope in places where it contacts rocks or sharp corners. Do not leave the rope knotted or stretched longer than necessary. Dry rope as soon as possible. Single-rope bridge construction is as follows:

Tie a sling rope around your waist using a square knot and two, separate half hitches. See the appendix for detailed information on knots.

Attach a locking steel carabiner to the sling rope.

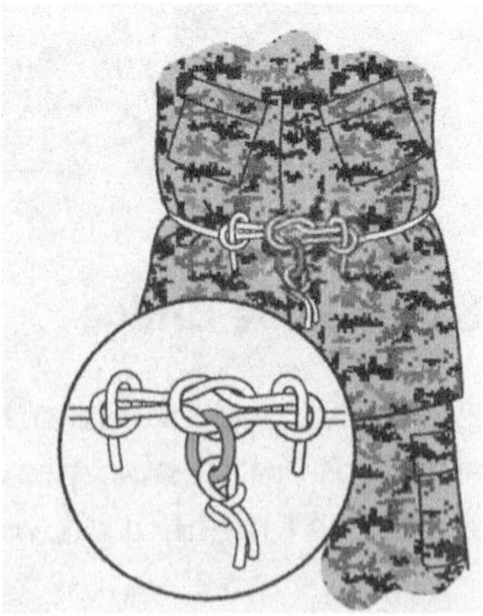

Tie a bowline knot in the running end of the bridge rope and attach it to the carabiner.

Temporarily secure the other end of the rope to a tree on the near shore.

Enter and cross the water.

> NOTE: Carry only your weapon and ammunition.

Exit the water on the opposite shore.

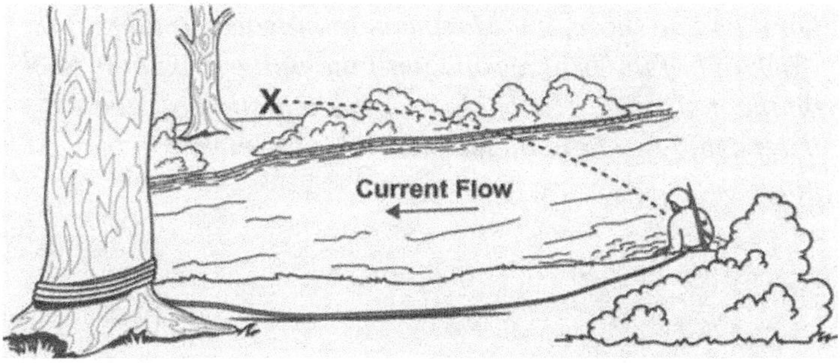

Prepare your weapon for use. Unhook the bridge rope from the carabiner at your waist, and tie the bridge rope to a sturdy tree using a round turn and two half hitches.

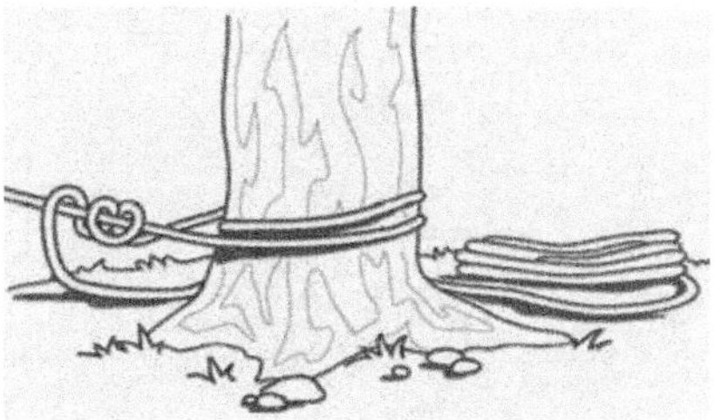

Conduct a box reconnaissance of the opposite shore.

On the near shore, have another Marine prepare to tighten the rope. That Marine should place a transport tightening system in the bridge rope by tying a double butterfly knot and placing two carabiners in the butterfly. See the appendix for detailed information on knots.

The Marine should pass the running end of the bridge rope around the downstream side of the near shore anchor point and through the two carabiners.

Pull the butterfly knot approximately one-third of the distance across the river.

Secure the bridge rope to an anchor point using a round turn and two half-hitches.

On the near shore, the Marine helping you should pull the slack out of the bridge rope until the butterfly knot is back on the near side. The bridge rope is then tied off against itself using two half hitches with a quick release in the last half hitch.

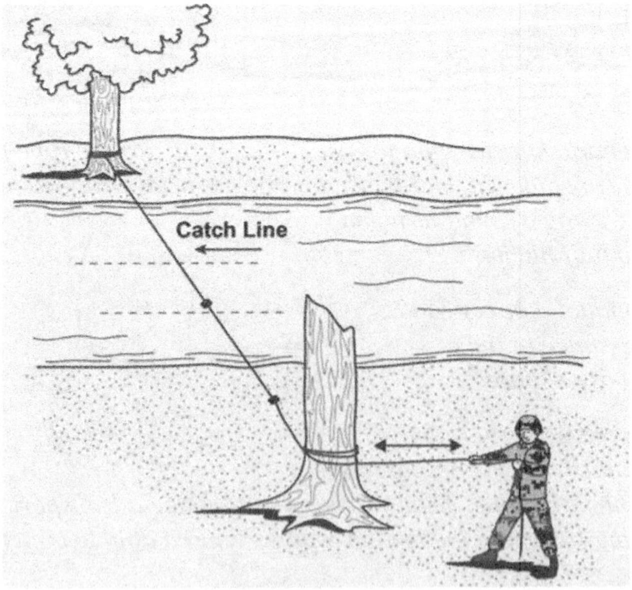

Note: The single-rope bridge must be as tight as possible so it will not sag when used.

If you lose your footing and fall into the water, swim with the current to the closest shore. Swimming against the current is dangerous and quickly causes fatigue.

High and Dry Crossings

If the single-rope bridge is high enough, suspend yourself below the single-rope bridge and above the water. Use the following steps to suspend yourself from a single-rope bridge and then pull yourself across the water:

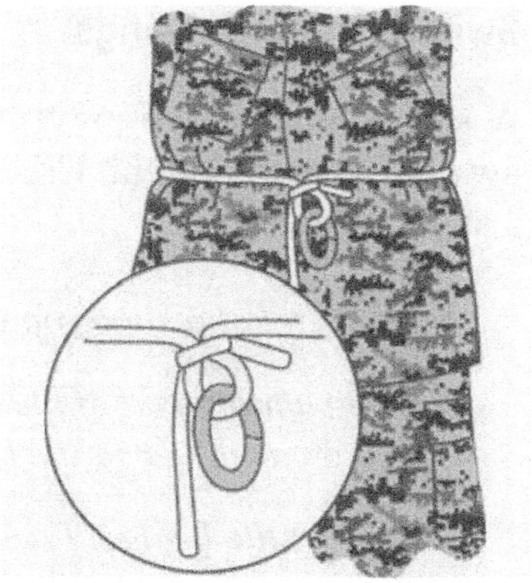

Tie a sling rope around your waist using a bowline. Ensure that the knot is tight.

Attach a carabiner through the bowline's loop. The carabiner's gate faces up.

Secure your helmet chin strap.

Face the single-rope bridge with your left shoulder toward the far shore.

Grasp the bridge rope in both hands.

Swing your body beneath the single-rope bridge with your head toward the far shore. Cross your ankles above the bridge rope.

Arch back until the carabiner contacts the bridge rope. Connect the carabiner to the bridge rope. Allow the carabiner to bear your body's weight.

Pull yourself across the single-rope bridge, hand over hand, to the far shore.

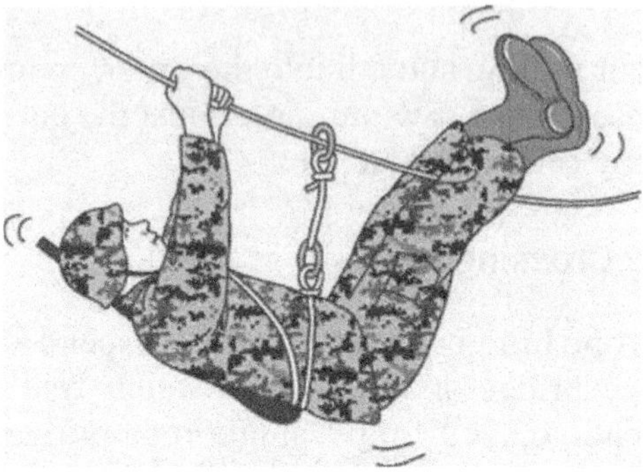

Swift Current Crossings

A single-rope bridge prevents being knocked down and swept away by a swift current. Use the following steps to move through a swift current:

Tie one end of a sling rope around your waist using a bowline.

Tie the running end of the sling rope in another bowline, and attach a carabiner to the bowline's loop.

Step up to the bridge. Face upstream.

Hook the carabiner to the single-rope bridge.

Walk sideways into the river while grasping the bridge rope in both hands.

Use the single-rope bridge for balance and remain standing, if possible.

Continue to move sideways through the river to the far shore.

Slow Current Crossings

If you face little or no current, it is not necessary to hook up to a bridge rope with a carabiner. Lie on your back in the water beneath the single-rope bridge. Support your body weight with your waterproof pack. Use the bridge rope and pull yourself hand over hand across the river.

Removal

If you are the last Marine waiting to cross, pull on the standing end of the rope to release the knot, then tie the rope around your waist using a bowline. The Marines on the far shore will pull you through the water.

Appendix
Knot Tying

Square Knot

The square knot is used to secure two ropes of equal diameter together so they form one continuous rope that will not slip.

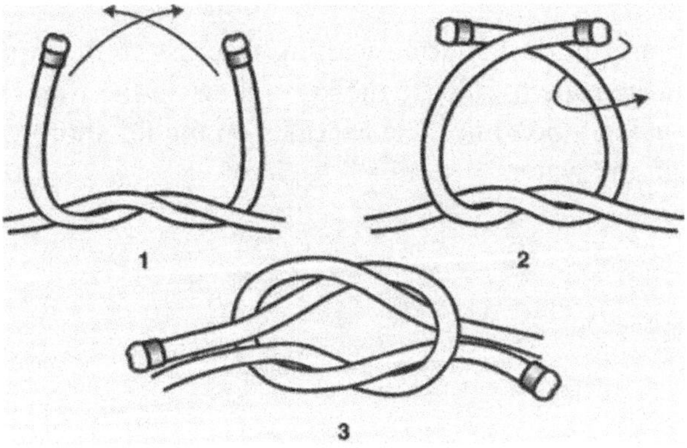

Bowline

The bowline forms a loop that will not tighten or slip under strain. It is easily untied.

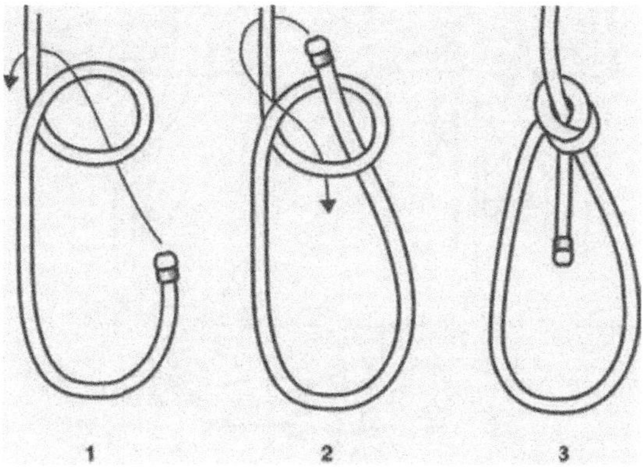

Hitches

Hitches are used to dress (prepare) knots and secure loose ends.

Half Hitch

The half hitch is used to tie a rope to a tree or to a larger rope. It will hold against a steady pull, but is not a secure hitch. It is frequently used to secure the free end of a rope.

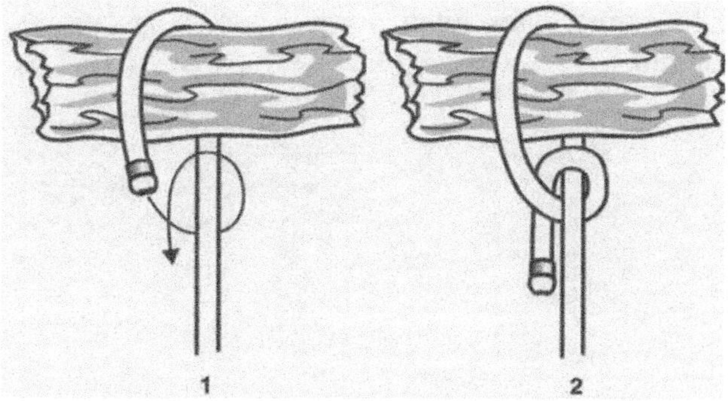

Two Half Hitches

Two half hitches can be used to secure the running end of a rope.

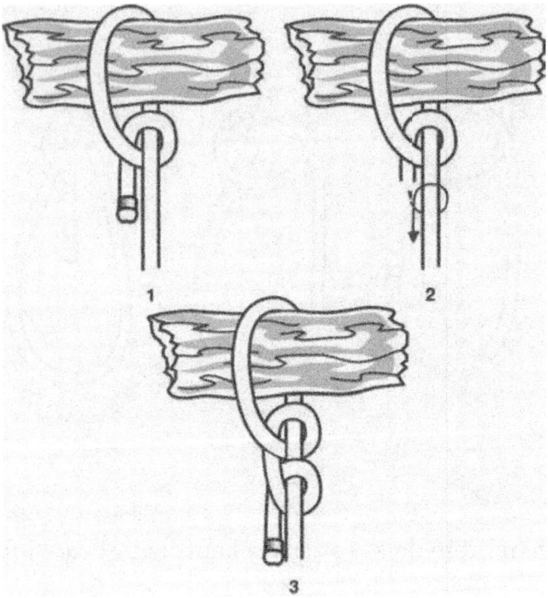

Round Turn and Two Half Hitches

A round turn and two half hitches can be used to fasten a rope to a tree. This hitch does not jam.

Butterfly Knot

The butterfly knot is used for anchor lines. Tied properly, it will not tighten on itself to the point that it cannot be easily untied. The butterfly knot can be used to pull a rope bridge taut. This knot can be used to tighten a fixed rope when mechanical means are not available. It will not jam if a stick is placed between the two upper loops.

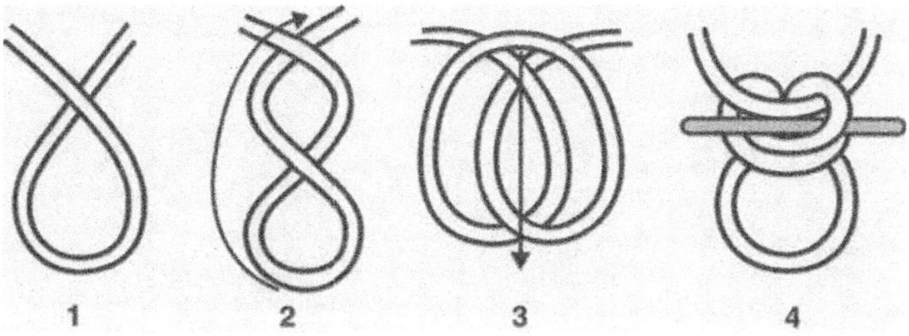

QUICK REFERENCE CHECKLIST

Decide to Survive!

S	-	**S**ize up the situation, surroundings, physical condition, equipment.
U	-	**U**se all your senses
R	-	**R**emember where you are.
V	-	**V**anquish fear and panic.
I	-	**I**mprovise and improve.
V	-	**V**alue living.
A	-	**A**ct like the natives.
L	-	**L**ive by your wits.

1. Immediate Actions

a. Assess immediate situation. ***THINK BEFORE YOU ACT!***

b. Take action to protect yourself from nuclear, biological, or chemical hazards (Chapter IX).

c. Seek a concealed site.

d. Assess medical condition; treat as necessary (Chapter V).

e. Sanitize uniform of potentially compromising information.

f. Sanitize area; hide equipment you are leaving.

g. Apply personal camouflage.

h. Move away from concealed site, zigzag pattern recommended.

i. Use terrain to advantage, communication, and concealment.

j. Find a hole-up site.

2. Hole-Up-Site (Chapter I)

a. Reassess situation; treat injuries, then inventory equipment.

b. Review plan of action; establish priorities (Chapter VI).

c. Determine current location.

d. Improve camouflage.

e. Focus thoughts on task(s) at hand.

f. Execute plan of action. Stay flexible!

Recommend inclusion of this manual in the aviator's survival vest.

3. Concealment (Chapter I)

a. Select a place of concealment providing—

(1) Adequate concealment, ground and air.

(2) Safe distance from enemy positions and lines of communications (LOC).

(3) Listening and observation points.

(4) Multiple avenues of escape.

(5) Protection from the environment.

(6) Possible communications/signaling opportunities.

b. Stay alert, maintain security.

c. Drink water.

4. Movement (Chapters I and II)

a. Travel slowly and deliberately.

b. ***DO NOT*** leave evidence of travel; use noise and light discipline.

c. Stay away from LOC.

d. Stop, look, listen, and smell; take appropriate action(s).

e. Move from one concealed area to another.

f. Use evasion movement techniques (Chapter I).

5. Communications and Signaling (Chapter III)

a. Communicate as directed in applicable plans/orders, particularly when considering transmitting ***in the blind***.

b. Be prepared to use communications and signaling devices on short notice.

c. Use of communications and signaling devices may compromise position.

6. Recovery (Chapter IV)

a. Select site(s) IAW criteria in theater recovery plans.

b. Ensure site is free of hazards; secure personal gear.

c. Select best area for communications and signaling devices.

d. Observe site for proximity to enemy activity and LOC.

e. Follow recovery force instructions.

FM 21-76-1
MCRP 3-02H
NWP 3-50.3
AFTTP(I) 3-2.26

FM 21-76-1	U.S. Army Training and Doctrine Command Fort Monroe, Virginia
MCRP 3-02H	Marine Corps Combat Development Command Quantico, Virginia
NWP 3-50.3	Navy Warfare Development Command Newport, Rhode Island
AFTTP(I) 3-2.26	Headquarters Air Force Doctrine Center Maxwell Air Force Base, Alabama

29 JUNE 1999

Survival, Evasion, and Recovery

Multiservice Procedures for Survival, Evasion, and Recovery

> **Note:** This *UNCLASSIFIED* publication is designed to provide Service members quick-reference survival, evasion, and recovery information. See Appendix B for the scope, purpose, application, implementation plan, and user information.

TABLE OF CONTENTS

THE CODE OF CONDUCT

ARTICLE I

I am an American, fighting in the forces which guard my country and our way of life. I am prepared to give my life in their defense.

ARTICLE II

I will never surrender of my own free will. If in command, I will never surrender the members of my command while they still have the means to resist.

ARTICLE III

If I am captured, I will continue to resist by all means available. I will make every effort to escape and aid others to escape. I will accept neither parole nor special favors from the enemy.

ARTICLE IV

If I become a prisoner of war, I will keep faith with my fellow prisoners. I will give no information or take part in any action which might be harmful to my comrades. If I am senior, I will take command. If not, I will obey the lawful orders of those appointed over me and will back them up in every way.

ARTICLE V

When questioned, should I become a prisoner of war, I am required to give name, rank, service number and date of birth. I will evade answering further questions to the utmost of my ability. I will make no oral or written statements disloyal to my country and its allies or harmful to their cause.

ARTICLE VI

I will never forget that I am an American, fighting for freedom, responsible for my actions, and dedicated to the principles which made my country free. I will trust in my God and in the United States of America.

Chapter I

EVASION

1. Planning

a. Review the quick reference checklist on the inside cover.

b. Guidelines for successful evasion include—

(1) Keeping a positive attitude.
(2) Using established procedures.
(3) Following your evasion plan of action.
(4) Being patient.
(5) Drinking water (***DO NOT*** eat food without water).
(6) Conserving strength for critical periods.
(7) Resting and sleeping as much as possible.
(8) Staying out of sight.

c. The following odors stand out and may give an evader away:

(1) Scented soaps and shampoos.
(2) Shaving cream, after-shave lotion, or other cosmetics.
(3) Insect repellent (camouflage stick is least scented).
(4) Gum and candy (smell is strong or sweet).
(5) Tobacco (odor is unmistakable).

d. Where to go (initiate evasion plan of action):

(1) Near a suitable area for recovery.
(2) Selected area for evasion.
(3) Neutral or friendly country or area.
(4) Designated area for recovery.

2. Camouflage

a. Basic principles:

(1) Disturb the area as little as possible.
(2) Avoid activity that reveals movement to the enemy.
(3) Apply personal camouflage.

b. Camouflage patterns **(Figure I-1)**:

(1) Blotch pattern.
  (a) Temperate deciduous (leaf shedding) areas.
  (b) Desert areas (barren).
  (c) Snow (barren).
(2) Slash pattern.
  (a) Coniferous areas (broad slashes).

(b) Jungle areas (broad slashes).
(c) Grass (narrow slashes).

(3) Combination. May use blotched and slash together.

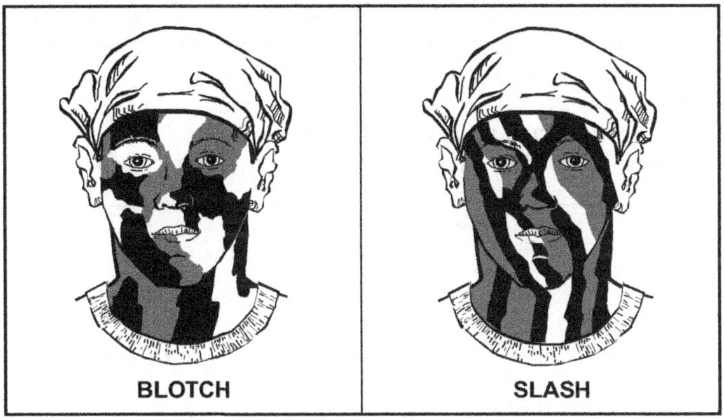

Figure I-1. Camouflage Patterns

c. Personal camouflage application follows:

(1) Face. Use dark colors on high spots and light colors on any remaining exposed areas. Use a hat, netting, or mask if available.

(2) Ears. The insides and the backs should have **2** colors to break up outlines.

(3) Head, neck, hands, and the under chin. Use scarf, collar, vegetation, netting, or coloration methods.

(4) Light colored hair. Give special attention to conceal with a scarf or mosquito head net.

d. Position and movement camouflage follows:

(1) Avoid unnecessary movement.

(2) Take advantage of natural concealment:

(a) Cut foliage fades and wilts, change regularly.
(b) Change camouflage depending on the surroundings.
(c) ***DO NOT*** select vegetation from same source.
(d) Use stains from grasses, berries, dirt, and charcoal.

(3) ***DO NOT*** over camouflage.

(4) Remember when using shadows, they shift with the sun.

(5) Never expose shiny objects (like a watch, glasses, or pens).

(6) Ensure watch alarms and hourly chimes are turned off.

(7) Remove unit patches, name tags, rank insignia, etc.

(8) Break up the outline of the body, ***"V"*** of crotch/armpits.

(9) Conduct observation from a prone and concealed position.

3. Shelters

a. Use camouflage and concealment.

b. Locate carefully—easy to remember acronym: ***BLISS***.

B - **B**lend **L** - **L**ow silhouette **I** - **I**rregular shape **S** - **S**mall **S** - **S**ecluded location

(1) Choose an area—

(a) Least likely to be searched (drainages, rough terrain, etc.) and blends with the environment.

(b) With escape routes (***DO NOT*** corner yourself).

(c) With observable approaches.

(2) Locate entrances and exits in brush and along ridges, ditches, and rocks to keep from forming paths to site.

(3) Be wary of flash floods in ravines and canyons.

(4) Conceal with minimal to no preparation.

(5) Take the direction finding threat into account before transmitting from shelter.

(6) Ensure overhead concealment.

4. Movement

a. A moving object is easy to spot. If travel is necessary—

(1) Mask with natural cover **(Figure I-2)**.

(2) Use the military crest.

(3) Restrict to periods of low light, bad weather, wind, or reduced enemy activity.

Figure I-2. Ground Movement

(4) Avoid silhouetting **(Figure I-3)**.

(5) At irregular intervals—

(a) ***STOP*** at a point of concealment.

(b) ***LOOK*** for signs of human or animal activity (smoke, tracks, roads, troops, vehicles, aircraft, wire, buildings, etc.). Watch for trip wires or booby traps and avoid leaving evidence of travel. Peripheral vision is more effective for recognizing movement at night and twilight.

(c) ***LISTEN*** for vehicles, troops, aircraft, weapons, animals, etc.

(d) ***SMELL*** for vehicles, troops, animals, fires, etc.

Figure I-3. Avoid Silhouetting

(6) Employ noise discipline; check clothing and equipment for items that could make noise during movement and secure them.

b. Break up the human shape or recognizable lines.

c. Route selection requires detailed planning and special techniques (irregular route/zigzag) to camouflage evidence of travel.

d. Some techniques for concealing evidence of travel follows:

(1) Avoid disturbing the vegetation above knee level.

(2) ***DO NOT*** break branches, leaves, or grass.

(3) Use a walking stick to part vegetation and push it back to its original position.

(4) ***DO NOT*** grab small trees or brush. (This may scuff the bark or create movement that is easily spotted. In snow country, this creates a path of snowless vegetation revealing your route.)

(5) Pick firm footing (carefully place the foot lightly but squarely on the surface to avoid slipping). ***TRY NOT TO***—

(a) Overturn ground cover, rocks, and sticks.

(b) Scuff bark on logs and sticks.

(c) Make noise by breaking sticks. (Cloth wrapped around feet helps muffle this.)

(d) Mangle grass and bushes that normally spring back.

(6) Mask unavoidable tracks in soft footing by—

(a) Placing tracks in the shadows of vegetation, downed logs, and snowdrifts.

(b) Moving before and during precipitation allows tracks to fill in.

(c) Traveling during windy periods.

(d) Taking advantage of solid surfaces (logs, rocks, etc.) leaving less evidence of travel.

(e) Patting out tracks lightly to speed their breakdown or make them look old.

(7) Secure trash or loose equipment—hide or bury discarded items. (Trash or lost equipment identifies who lost it.)

(8) Concentrate on defeating the handler if pursued by dogs.

e. Penetrate obstacles as follows:

(1) Enter deep ditches feet first to avoid injury.

(2) Go around chain-link and wire fences. Go under fence if unavoidable, crossing at damaged areas. ***DO NOT*** touch fence; look for electrical insulators or security devices.

(3) Penetrate rail fences, passing under or between lower rails. If impractical, go over the top, presenting as low a silhouette as possible **(Figure I-4)**.

(4) Cross roads after observation from concealment to determine enemy activity. Cross at points offering concealment such as bushes, shadows, bend in road, etc. Cross in a manner leaving your footprints parallel (cross step sideways) to the road. **(Figure I-5)**

(5) Use same method of observation for railroad tracks that was used for roads. Next, align body parallel to tracks with face down, cross tracks using a semi-pushup motion. Repeat for the second track. **(Figure I-6)**.

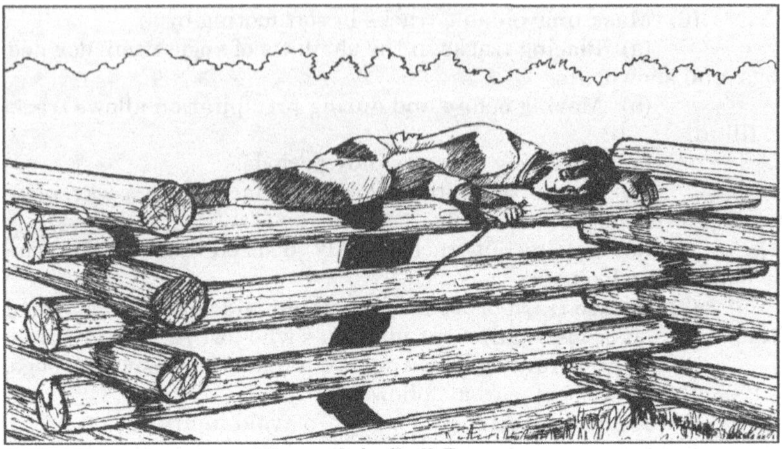

Figure I-4. Rail Fences

Figure I-5. Road Crossing

Figure I-6. Railroad Tracks

WARNING: If 3 rails exist, 1 may be electrified.

Chapter II
NAVIGATION

Assess the threat and apply appropriate evasion principles.

1. Stay or Move Considerations

a. Stay with the vehicle/aircraft in a non-combat environment.

b. Leave only when—

(1) Dictated by the threat.

(2) Are certain of your location, have a known destination, and have the ability to get there.

(3) Can reach water, food, shelter, and/or help.

(4) Convinced rescue is not coming.

c. Consider the following if you decide to travel:

(1) Follow the briefed evasion plan.

(2) Determine which direction to travel and why.

(3) Decide what equipment to take, cache, or destroy.

d. Leave information at your starting point (in a non-combat environment) that includes—

(1) Destination.

(2) Route of travel.

(3) Personal condition.

(4) Supplies available.

e. Consider the following for maps (in a combat environment):

(1) ***DO NOT*** write on the map.

(2) ***DO NOT*** soil the map by touching the destination.

(3) ***DO NOT*** fold in a manner providing travel information.

Note: **These actions may compromise information if captured.**

2. Navigation and Position Determination

a. Determine your general location by—

(1) Developing a working knowledge of the operational area.

(a) Geographic checkpoints.

(b) Man-made checkpoints.

(c) Previous knowledge of operational area.

(2) Using the ***Rate x Time = Distance*** formula.

(3) Using information provided in the map legend.

(4) Using prominent landmarks.

(5) Visualizing map to determine position.

b. Determine cardinal directions (north, south, east, and west) by—

(1) Using compass.

CAUTION: The following methods are ***NOT*** highly accurate and give only general cardinal direction.

(2) Using stick and shadow method to determine a true north-south line **(Figure II-1)**.

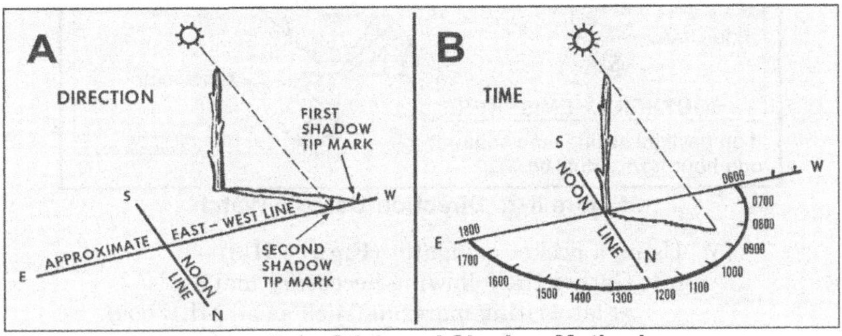

Figure II-1. Stick and Shadow Method

(3) Remembering the sunrise/moonrise is in the east and sunset/moonset is in the west.

(4) Using a wristwatch to determine general cardinal direction **(Figure II-2)**.

(a) Digital watches. Visualize a clock face on the watch.

(b) Northern Hemisphere. Point hour hand at the sun. South is halfway between the hour hand and 12 o'clock position.

(c) Southern Hemisphere. Point the 12 o'clock position on your watch at the sun. North is halfway between the 12 o'clock position and the hour hand.

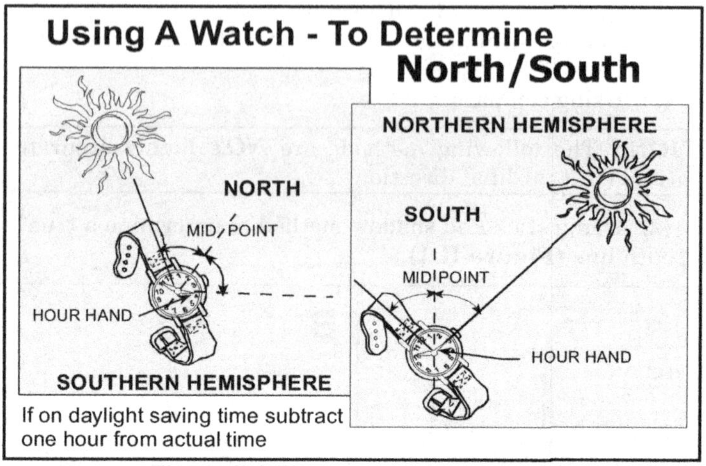

Figure II-2. Direction Using a Watch

(5) Using a pocket navigator **(Figure II-3)**—

(a) Gather the following necessary materials:

•Flat writing material (such as an MRE box).

•1-2 inch shadow tip device (a twig, nail, or match).

•Pen or pencil.

(b) Start construction at sunup; end construction at sundown. Do the following:

•Attach shadow tip device in center of paper.

•Secure navigator on flat surface (***DO NOT*** move during set up period).

•Mark tip of shadow every 30 minutes annotating the time.

•Connect marks to form an arc.

•Indicate north with a drawn arrow.

Note: The shortest line between base of shadow tip device and curved line is a north-south line.

(c) Do the following during travel:

•Hold navigator so the shadow aligns with mark of present time (drawn arrow now points to true north).

(d) Remember the navigator is current for approximately 1 week.

CAUTION: The Pocket Navigator is ***NOT*** recommended if evading.

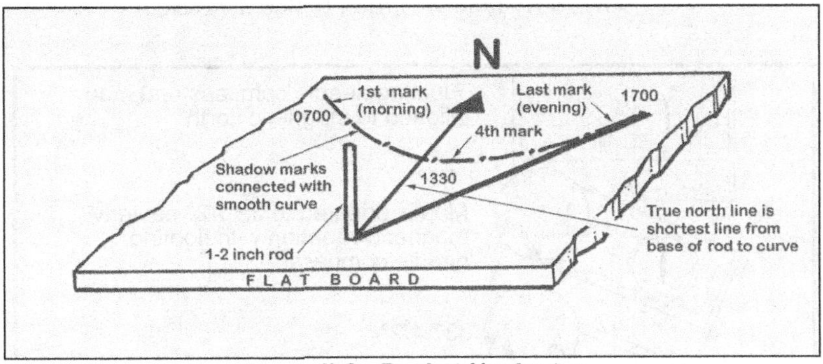

Figure II-3. Pocket Navigator

(6) Using the stars **(Figure II-4)** the—

(a) North Star is used to locate true north-south line.

(b) Southern Cross is used to locate true south-north line.

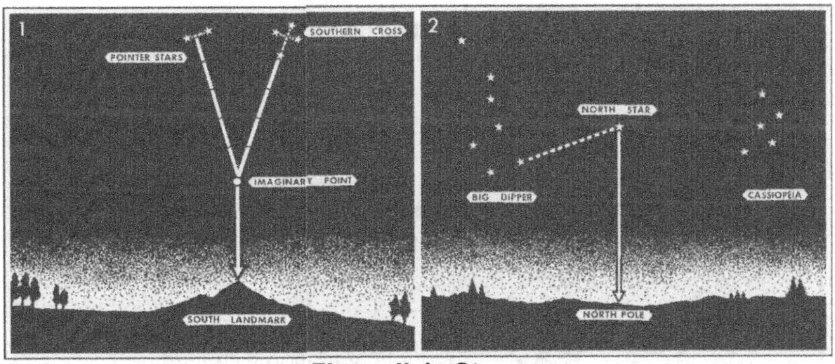

Figure II-4. Stars

c. Orient the map by—

(1) Using a true north-south line **(Figure II-5)**—

(a) Unfold map and place on a firm, flat, level nonmetallic surface.

(b) Align the compass on a true north-south line.

(c) Rotate map and compass until stationary index line aligns with the magnetic variation indicated in marginal information.

- Easterly (subtract variation from 360 degrees).
- Westerly (add variation to 360 degrees).

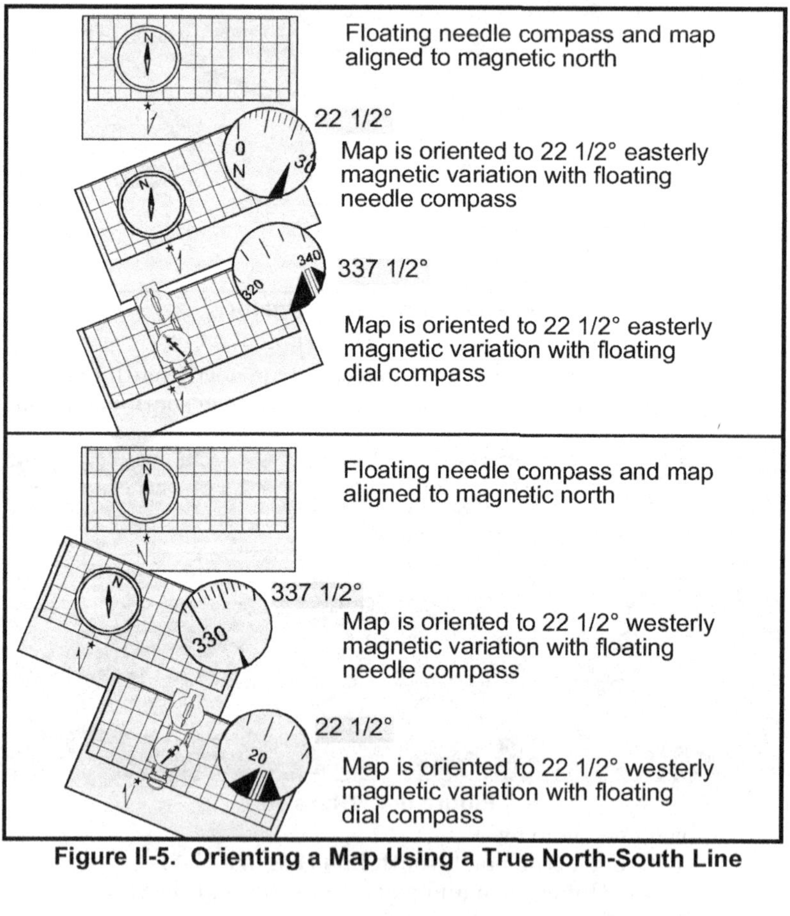

Figure II-5. Orienting a Map Using a True North-South Line

(2) Using a compass rose **(Figure II-6)**—

(a) Place edge of the lensatic compass on magnetic north line of the compass rose closest to your location.

(b) Rotate map and compass until compass reads 360 degrees.

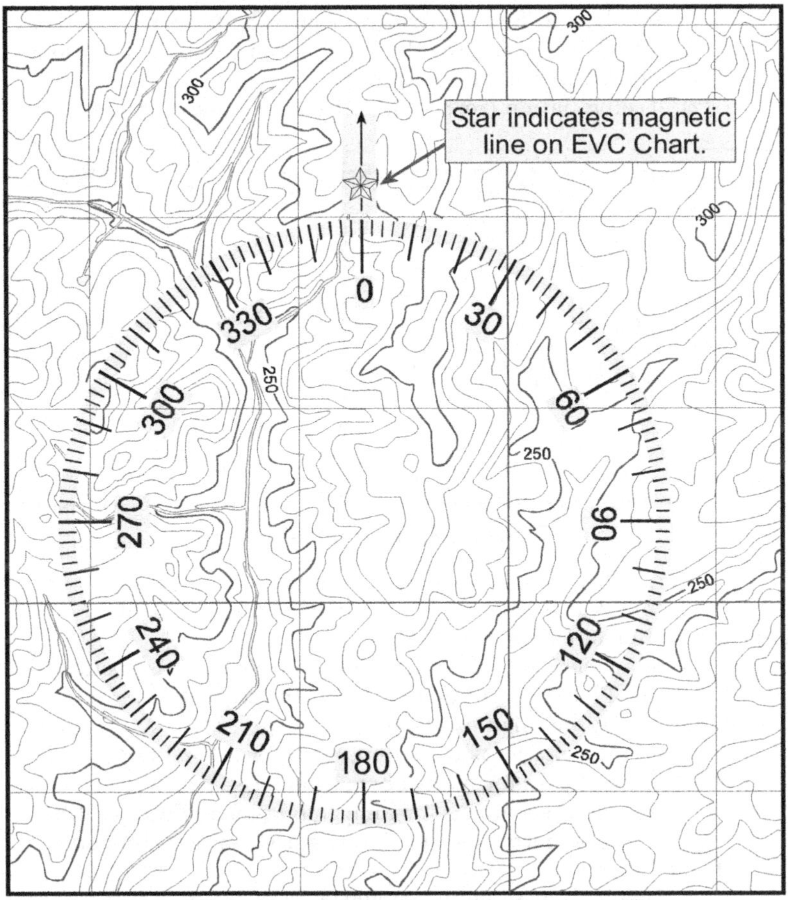

Figure II-6. Map Orientation with Compass Rose

(3) If there is **NO** compass, orient map using cardinal direction obtained by the stick and shadow method or the celestial aids (stars) method.

d. Determine specific location.

(1) Global Positioning System (GPS).

(a) ***DO NOT*** use GPS for primary navigation.

(b) Use GPS to confirm your position ***ONLY***.

(c) Select area providing maximum satellite reception.

(d) Conserve GPS battery life.

(2) Triangulation (resection) with a compass **(Figure II-7)**.

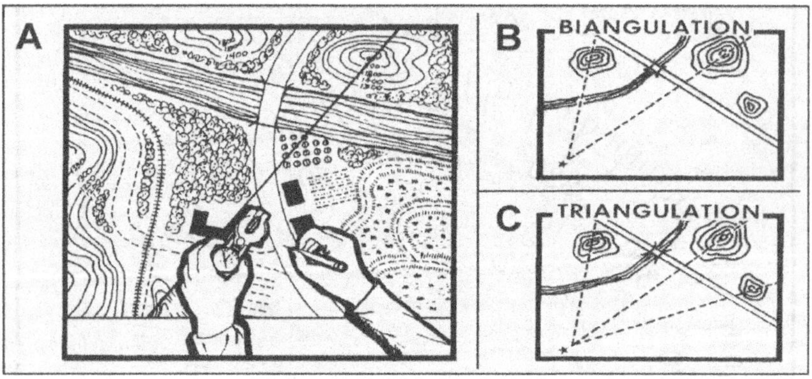

Figure II-7. Triangulation

(a) Try to use **3** or more azimuths.

(b) Positively identify a major land feature and determine a line of position (LOP).

(c) Check map orientation each time compass is used.

(d) Plot the LOP using a thin stick or blade of grass (combat) or pencil line (non-combat).

(e) Repeat steps **(b)** through **(d)** for other LOPs.

e. Use the compass for night navigation by—

(1) Setting up compass for night navigation **(Figure II-8)**.

(2) Aligning north-seeking arrow with luminous line and follow front of compass.

(3) Using point-to-point navigation.

f. Route selection techniques follow:

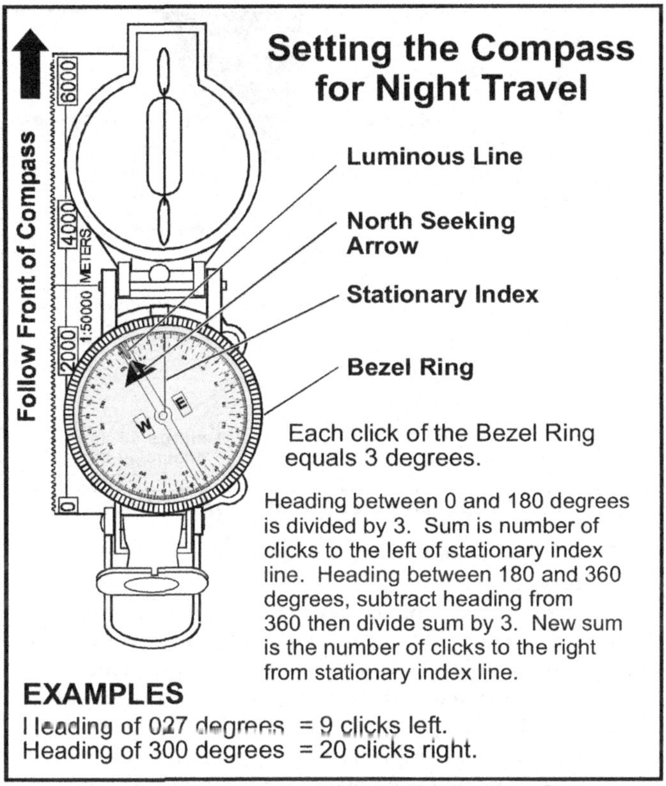

Figure II-8. Compass Night Navigation Setup

(1) Circumnavigation.

(a) Find a prominent landmark on the opposite side of the obstacle.

(b) Contour around obstacle to landmark.

(c) Resume your route of travel.

(2) Dogleg and 90 degree offset **(Figure II-9)**.

(3) Straight-line heading as follows:

(a) Maintain heading until reaching destination.

(b) Measure distance by counting the number of paces in a given course and convert to map units.

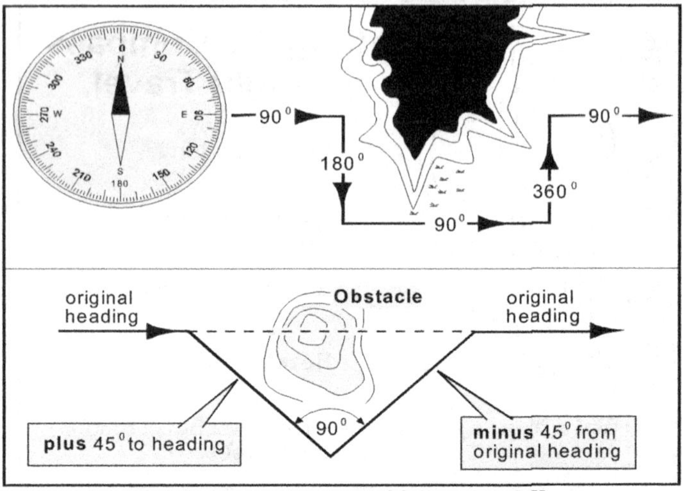

Figure II-9. Dogleg and 90 Degree Offset

•One pace is the distance covered each time the same foot touches the ground.

•Distances measured by paces are approximate (example in open terrain, 900 paces per kilometer [average], or example in rough terrain, 1200 paces per kilometer [average]).

(c) Use pace count in conjunction with terrain evaluation and heading to determine location. An individual's pace varies because of factors such as steep terrain, day/night travel, or injured/uninjured condition. Adjust estimation of distance traveled against these factors to get relative accuracy when using a pace count.

(4) Deliberate offset is—

(a) Used when finding a point on a linear feature (that is, road or river).

(b) Intentionally navigated to left or right of target so you know which way to turn at the linear feature.

(5) Point-to-point is same as straight line.

(a) Pick out landmarks on the heading and walk the trail of least resistance to a point.

(b) On reaching a point, establish another landmark and continue.

3. Travel Considerations

a. Pick the easiest and safest route (non-combat).

b. Maintain a realistic pace; take rest stops when needed.

c. Avoid overdressing and overheating.

d. Consider food and water requirements.

e. Take special care of feet (change socks regularly).

f. Pack equipment to prevent loss, damage, pack imbalance, and personal safety.

g. Go **around** obstacles, not over or through them.

h. Travel on trails whenever possible (non-combat).

i. Travel in forested areas if possible.

j. Avoid creek bottoms and ravines with ***NO*** escape in the event of heavy rains.

k. Consider the following for swamps, lakes, and unfordable rivers:

(1) Circumnavigate swamps, lakes, and bogs if needed.

(2) Travel downstream to find people and slower water.

(3) Travel upstream to find narrower and shallow water.

4. River Travel

River travel may be faster and save energy when hypothermia is not a factor. It may be a primary mode of travel and LOC in a tropical environment **(use with caution if evading)**.

a. Use flotation device (raft, log, bamboo, etc.).

b. Use a pole to move the raft in shallow water.

c. Use an oar in deep water.

d. Stay near inside edge of river bends (current speed is less).

e. Keep near shore.

f. Watch for the following ***DANGERS:***

(1) Snags.

(2) Sweepers (overhanging limbs and trees).

(3) Rapids (***DO NOT*** attempt to shoot the rapids).

(4) Waterfalls.

(5) Hazardous animals.

g. Consider using a flotation device when crossing rivers or large/deep streams.

5. Ice and Snow Travel

Travel should be limited to areas free of hazards.

a. ***DO NOT*** travel in—

(1) Blizzards.

(2) Bitterly cold winds.

(3) Poor visibility.

b. Obstacles to winter travel follow:

(1) Reduced daylight hours (***BE AWARE***).

(2) Deep soft snow (if movement is necessary, make snowshoes **[Figure II-10])**. Travel is easier in early morning or late afternoon near dusk when snow is frozen or crusted.

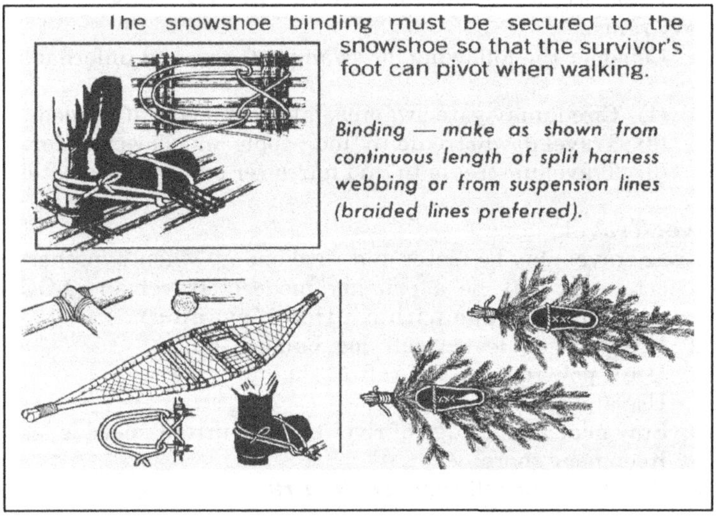

Figure II-10. Improvised Snowshoes

(3) Avalanche prone areas to avoid:

(a) Slopes 30-45 degrees or greater.

(b) Trees without uphill branches (identifies prior avalanches).

(c) Heavy snow loading on ridge tops.

(4) If caught in an avalanche, do the following:

(a) Backstroke to decrease burial depth.

(b) Move hand around face to create air pocket as moving snow slows.

(5) Frozen water crossings.

(a) Weak ice should be expected where—

- Rivers are straight.
- Objects protrude through ice.
- Snow banks extend over the ice.
- Rivers or streams come together.
- Water vapor rising indicates open or warm areas.

(b) Air pockets form when a frozen river loses volume.

(c) When crossing frozen water, distribute your weight by laying flat, belly crawling, or using snowshoes.

c. Glacier travel is hazardous and should be avoided.

6. Mountain Hazards

a. Lightning. Avoid ridge tops during thunderstorms.

b. Avalanche. Avoid areas prone to avalanches.

c. Flash floods. Avoid low areas.

7. Summer Hazards (see page II-10; paragraph 3, *Travel Considerations*, items h through k.)

(1) Dense brush.

(a) Travel on trails when possible (non-combat).

(b) Travel in forested areas if possible.

(c) Avoid creek bottoms and ravines with no escape in the event of heavy rains.

(2) Swamps, lakes, and unfordable rivers.

(a) Circumnavigate swamps, lakes, and bogs if needed.

(b) Travel downstream to find people and slower water.

(c) Travel upstream to find narrower and shallow water.

8. Dry Climates

a. ***DO NOT*** travel unless certain of reaching the destination using the water supply available.

b. Travel at dawn or dusk on hot days.

c. Follow the easiest trail possible (non-combat), avoiding—

(1) Deep sandy dune areas.

(2) Rough terrain.

d. In sand dune areas—

(1) Follow hard valley floor between dunes.

(2) Travel on the windward side of dune ridges.

e. If a sandstorm occurs—

(1) Mark your direction of travel.

(2) Sit or lie down in direction of travel.

(3) Try to get to the downwind side of natural shelter.

(4) Cover the mouth and nose with a piece of cloth.

(5) Protect the eyes.

(6) Remain stationary until the storm is over.

9. Tropical Climates

a. Travel only when it is light.

b. Avoid obstacles like thickets and swamps.

c. Part the vegetation to pass through. Avoid grabbing vegetation; it may have spines or thorns (**use gloves** if possible).

d. ***DO NOT*** climb over logs if you can go around them.

e. Find trails—

(1) Where **2** streams meet.

(2) Where a low pass goes over a range of hills.

f. While traveling trails—

(1) Watch for disturbed areas on game trails; they may indicate a pitfall or trap.

(2) Use a walking stick to probe for pitfalls or traps.

(3) ***DO NOT*** sleep on the trail.

(4) Exercise caution, the enemy uses the trails also.

10. Open Seas

a. Using currents—

(1) Deploy sea anchor **(Figure II-11)**. Sea anchor may be adjusted to make use of existing currents.

(2) Sit low in the raft.

(3) Deflate the raft slightly so it rides lower in the water.

b. Using winds—

(1) Pull in sea anchor.

(2) Inflate raft so it rides higher.

(3) Sit up in raft so body catches the wind.

(4) Construct a shade cover/sail **(Figure II-12)**. (Sail aids in making landfall.)

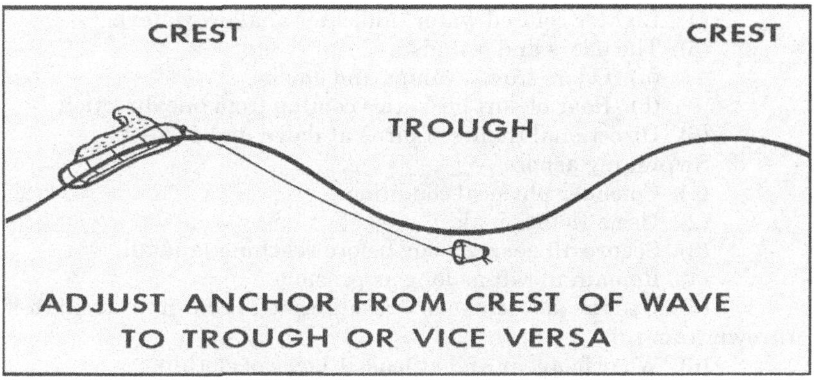

Figure II-11. Sea Anchor Deployment

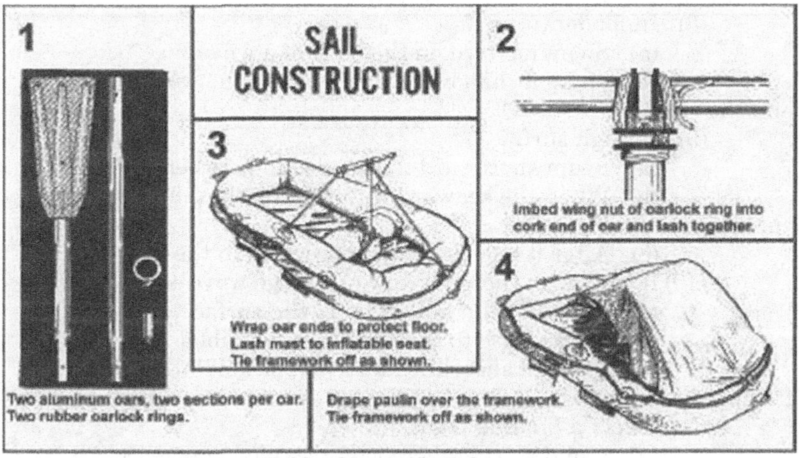

Figure II-12. Shade/Sail Construction

c. Making landfall. Indications of land are—

(1) Fixed cumulus clouds in a clear sky or in a cloudy sky where all other clouds are moving.

(2) Greenish tint in the sky **(in the tropics)**.

(3) Lighter colored reflection on clouds (open water causes dark gray reflections) **(in the arctic)**.

(4) Lighter colored water (indicates shallow water).

(5) The odors and sounds.

(a) Odors from swamps and smoke.

(b) Roar of surf/bird cries coming from one direction.

(6) Directional flights of birds at dawn and at dusk.

d. Swimming ashore—

(1) Consider physical condition.

(2) Use a flotation aid.

(3) Secure all gear to body before reaching landfall.

(4) Remain in raft as long as possible.

(5) Use the sidestroke or breaststroke to conserve strength if thrown from raft.

(6) Wear footgear and at least **1** layer of clothing.

(7) Try to make landfall during the lull between the sets of waves (waves are generally in **sets** of **7**, from **smallest** to **largest**).

(8) In moderate surf.

(a) Swim forward on the back of a wave.

(b) Make a shallow dive just before the wave breaks to end the ride.

(9) In high surf.

(a) Swim shoreward in the trough between waves.

(b) When the seaward wave approaches, face it and submerge.

(c) After it passes, work shoreward in the next trough.

(10) If caught in the undertow of a large wave—

(a) Remain calm and swim to the surface.

(b) Lie as close to the surface as possible.

(c) Parallel shoreline and attempt landfall at a point further down shore.

(11) Select a landing point.

(a) Avoid places where waves explode upon rocks.

(b) Find a place where waves smoothly rush onto the rocks.

(12) After selecting a landing site—

(a) Face shoreward.

(b) Assume a sitting position with feet 2 or 3 feet lower than head to absorb the shock of hitting submerged objects.

e. Rafting ashore—

(1) Select landing point carefully.

(2) Use caution landing when the sun is low and straight in front of you causing poor visibility.

(3) Land on the lee (downwind) side of islands or point of land if possible.

(4) Head for gaps in the surf line.

(5) Penetrate surf by—

(a) Taking down most shade/sails.

(b) Using paddles to maintain control.

(c) Deploying a sea anchor for stability.

CAUTION: ***DO NOT*** deploy a sea anchor if traveling through coral.

f. Making sea ice landings on large stable ice flows. Icebergs, small flows, and disintegrating flows are dangerous **(ice can cut a raft)**.

(1) Use paddles to avoid sharp edges.

(2) Store raft away from the ice edge.

(3) Keep raft inflated and ready for use.

(4) Weight down/secure raft so it does not blow away.

Chapter III

RADIO COMMUNICATIONS AND SIGNALING

Inventory and review the operating instructions of all communications and signaling equipment.

1. Radio Communications (Voice and Data)

a. Non-combat.

(1) Ensure locator beacon is operational.

(2) Follow standing plans for on/off operations to conserve battery use.

b. Combat.

(1) Turn off locator beacon.

(2) Keep it with you to supplement radio communications.

(3) Follow plans/orders for on/off operations.

c. Make initial contact as soon as possible or as directed in applicable plans/orders.

d. If no immediate contact, then as directed in applicable plans/orders.

e. Locate spare radio and batteries (keep warm and dry).

f. Transmissions.

(1) Use concealment sites (combat) that optimize line of site (LOS).

(2) Face recovery asset.

(3) Keep antenna perpendicular to intended receiver **(Figure III-1)**.

(4) ***DO NOT*** ground antenna (that is finger on antenna or attaching bolt, space blanket, vegetation, etc.).

(5) Keep transmissions short (3-5 seconds maximum). Use data burst if available.

(6) Move after each transmission (***ONLY*** in combat, if possible).

(7) If transmitting in the blind, ensure a clear LOS towards the equator.

(8) Use terrain masking to hinder enemy direction finding.

g. Listening (use reception times in applicable plans/orders or as directed by recovery forces).

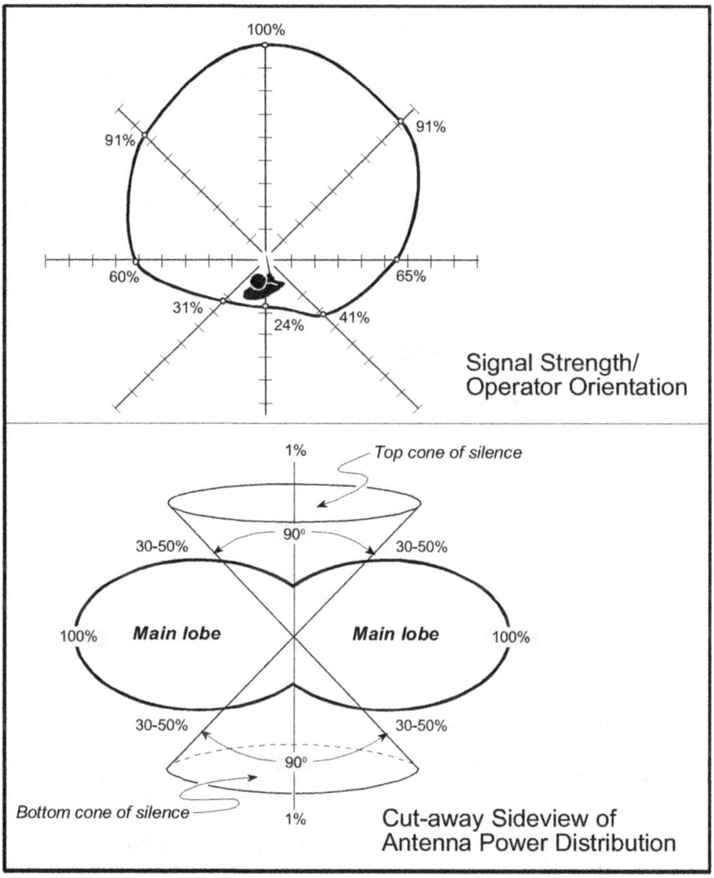

Figure III-1. Radio Transmission Characteristics

2. Signaling

a. Pyrotechnic signals.

(1) Prepare early (weather permitting).

(2) Use as directed in applicable plans/orders or as directed by recovery forces.

(3) Extend over raft's edge before activating.

b. Signal mirror **(Figure III-2)**.
   (1) Use as directed by recovery forces.
   (2) If no radio, use only with confirmed friendly forces.
   (3) Cover when not in use.

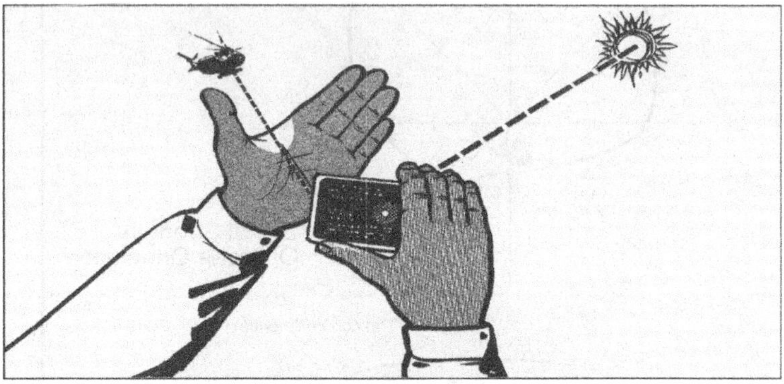

Figure III-2. Sighting Techniques

Note: Make a mirror from any shiny metal or glass.

c. Strobe/IR lights.
   (1) Prepare early, consider filters and shields.
   (2) Use as directed by recovery forces.
   (3) Conserve battery life.

Note: Produces one residual flash when turned off.

d. Pattern signals (use as directed in applicable plans/orders).
   (1) Materials:
      (a) Manmade (space blanket, signal paulin, parachute).
      (b) Natural use materials that contrast the color and/or texture of the signaling area (rocks, brush, branches, stomped grass).
   (2) Location.
      (a) Maximize visibility from above.
      (b) Provide concealment from ground observation.
   (3) Size (large as possible) and ratio **(Figure III-3)**.

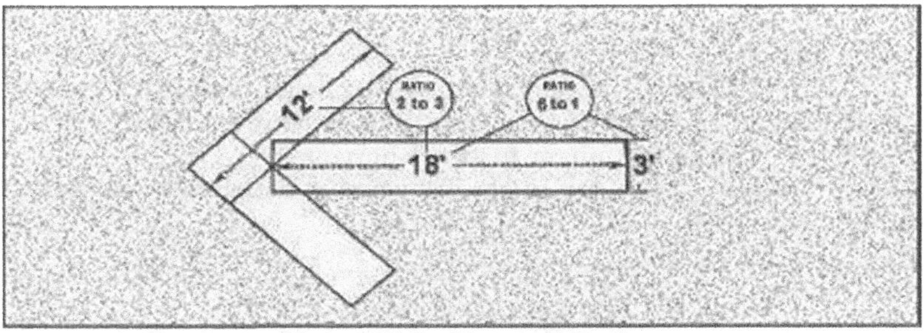

Figure III-3. Size and Ratio

(4) Shape (maintain straight lines and sharp corners).
(5) Contrast (use color and shadows).
(6) Pattern signals **(Figure III-4)**.

NO.	MESSAGE	CODE SYMBOL
1	REQUIRE ASSISTANCE	V
2	REQUIRE MEDICAL ASSISTANCE	X
3	NO or NEGATIVE	N
4	YES or AFFIRMATIVE	Y
5	PROCEEDING IN THIS DIRECTION	↑

Figure III-4. Signal Key

e. Sea dye marker.
(1) ***DO NOT*** waste in rough seas or fast moving water.
(2) Conserve unused dye by rewrapping.
(3) May be used to color snow.

f. Non-combat considerations:

(1) Use a fire at night.

(2) Use smoke for day (tires or petroleum products for dark smoke and green vegetation for light smoke). **(Figure III-5)**

(3) Use signal mirror to sweep horizon.

(4) Use audio signals (that is, voice, whistle, and weapons fire).

Figure III-5. Smoke Generator

Chapter IV
RECOVERY

1. Responsibilities

a. Establish radio contact with recovery forces (if possible).

b. Maintain communication with recovery forces until recovered.

c. Be prepared to authenticate as directed in applicable plans/orders.

d. Follow recovery force instructions, be prepared to report—

(1) Enemy activity in the recovery area.

(2) Recovery site characteristics (slope, obstacles, size, etc.).

(3) Number in party/medical situation.

(4) Signal devices available.

e. If no radio, a ground-to-air signal may be your only means to effect recovery.

2. Site Selection

a. Locate area for landing pick-up, if practical (approximately 150 feet diameter, free of obstructions, flat and level).

b. Assess evidence of human activity at/near the site (in combat).

c. Locate several concealment sites around area (in combat).

d. Plan several tactical entry and exit routes (in combat).

3. Site Preparation

a. Pack and secure all equipment.

b. Prepare signaling devices (use as directed or as briefed).

c. Mentally review recovery methods (aircraft, ground, boat, etc.).

4. Recovery Procedures

a. Assist recovery force in identifying your position.

b. Stay concealed until recovery is imminent (in combat).

c. For a landing/ground recovery—

(1) Assume a non-threatening posture.

(2) Secure weapons and avoid quick movement.

(3) ***DO NOT*** approach recovery vehicle until instructed.

(4) Beware of rotors/propellers when approaching recovery vehicle, especially on sloping or uneven terrain. Secure loose equipment that could be caught in rotors/propellers.

d. For **hoist** recovery devices **(Figures IV-1 and IV-2)**—

(1) Use eye protection, if available (glasses or helmet visor).

(2) Allow metal on device to contact the surface before touching to avoid injury from static discharge.

(3) Sit or kneel for stability while donning device.

(4) Put safety strap under armpits.

(5) Ensure cable is in front of you.

(6) Keep hands clear of all hardware and connectors.

(7) ***DO NOT*** become entangled in cable.

(8) Use a thumbs up, vigorous cable shake, or radio call to signal you are ready.

(9) Drag feet on the ground to decrease oscillation.

(10) ***DO NOT*** assist during hoist or when pulled into the rescue vehicle. Follow crewmember instructions.

e. For **nonhoist** recovery (rope or unfamiliar equipment)—

(1) Create a ***"fixed loop"*** big enough to place under armpits **(Figure IV-3)**.

(2) Follow the procedures in **"d"** above.

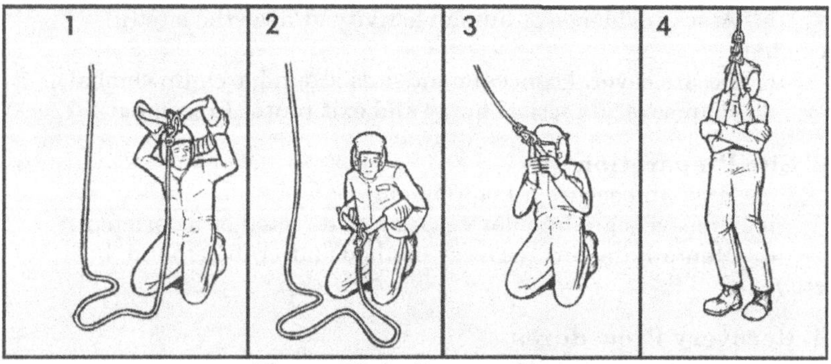

Figure IV–1. Rescue Strap

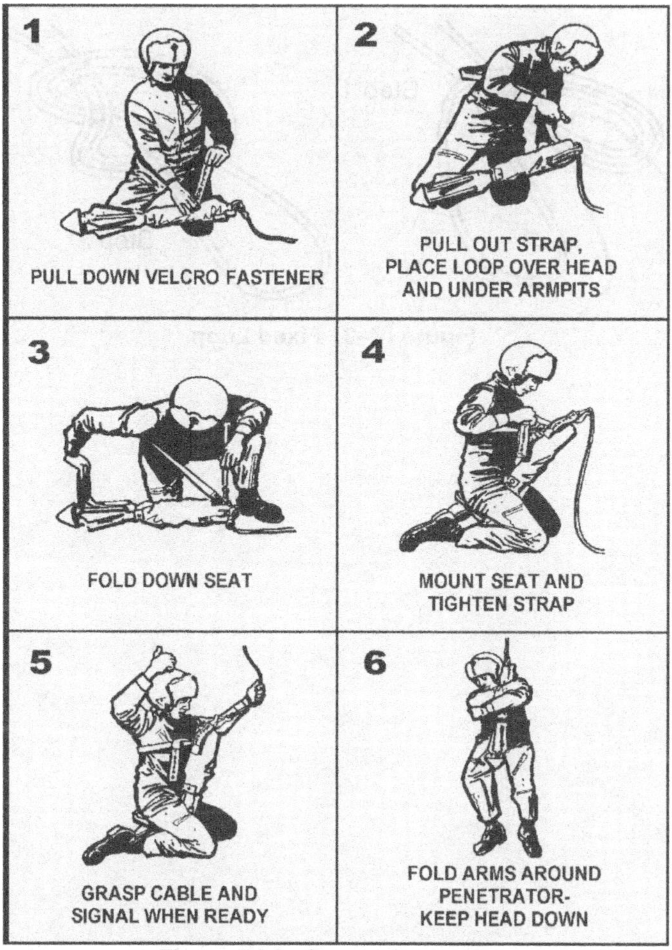

Figure IV-2. Forest Penetrator

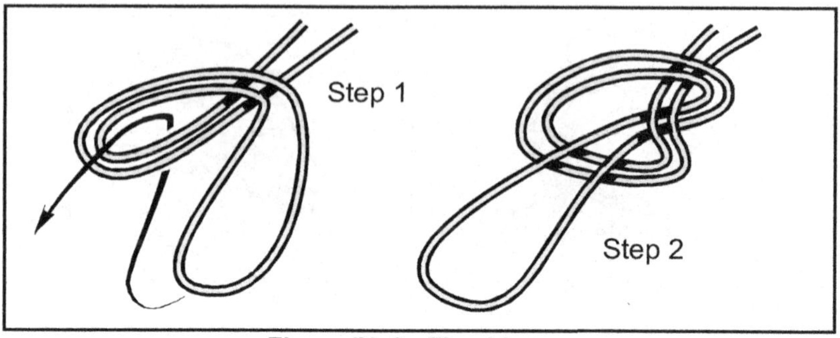

Figure IV–3. Fixed Loop

Chapter V
MEDICAL

WARNING: These emergency medical procedures are for survival situations. Obtain professional medical treatment as soon as possible.

1. Immediate First Aid Actions

Remember the *ABCs* of Emergency Care:
Airway Breathing Circulation

a. Determine responsiveness as follows:
   (1) If unconscious, arouse by shaking gently and shouting.
   (2) If no response—
      (a) Keep head and neck aligned with body.
      (b) Roll victims onto their backs.
      (c) Open the airway by lifting the chin **(Figure V-1)**.
      (d) Look, listen, and feel for air exchange.

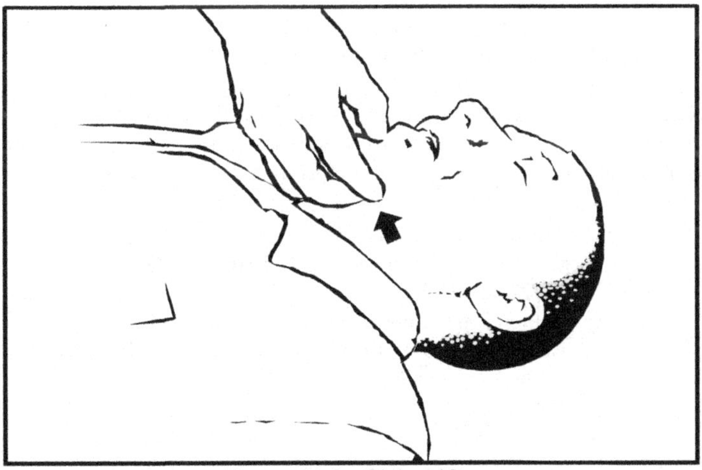

Figure V-1. Chin Lift

(3) If victim is not breathing—

(a) Check for a clear airway; remove any blockage.

(b) Cover victim's mouth with your own.

(c) Pinch victim's nostrils closed.

(d) Fill victim's lungs with **2** slow breaths.

(e) If breaths are blocked, reposition airway; try again.

(f) If breaths still blocked, give **5** abdominal thrusts:

•Straddle the victim.

•Place a fist between breastbone and belly button.

•Thrust upward to expel air from stomach.

(g) Sweep with finger to clear mouth.

(h) Try **2** slow breaths again.

(i) If the airway is still blocked, continue **(c)** through **(f)** until successful or exhausted.

(j) With open airway, start mouth to mouth breathing:

•Give **1** breath every 5 seconds.

•Check for chest rise each time.

(4) If victim is unconscious, but breathing—

(a) Keep head and neck aligned with body.

(b) Roll victim on side (drains the mouth and prevents the tongue from blocking airway).

(5) If breathing difficulty is caused by chest trauma, refer to **page V-7, paragraph 1d, *Treat Chest Injuries***.

CAUTION: ***DO NOT*** remove an impaled object unless it interferes with the airway. You may cause more tissue damage and increase bleeding. For travel, you may shorten and secure the object.

b. Control bleeding as follows:

(1) Apply a pressure dressing **(Figure V-2)**.

(2) If ***STILL*** bleeding—

(a) Use direct pressure over the wound.

(b) Elevate the wounded area above the heart.

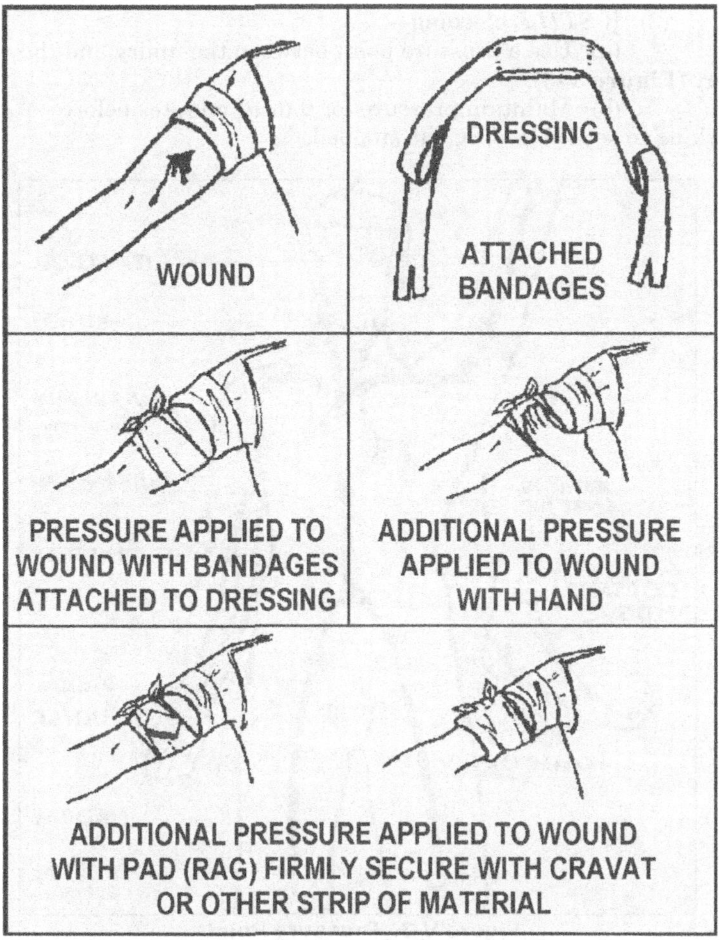

Figure V-2. Application of a Pressure Dressing

(3) If ***STILL*** bleeding—

(a) Use a pressure point between the injury and the heart **(Figure V-3)**.

(b) Maintain pressure for 6 to 10 minutes before checking to see if bleeding has stopped.

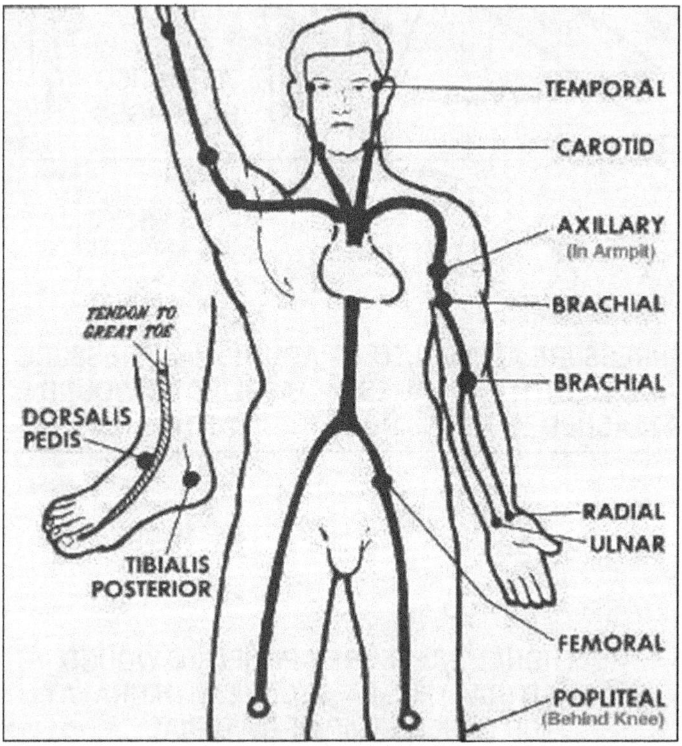

Figure V-3. Pressure Points

(4) If a limb wound is ***STILL*** bleeding—

CAUTION: Use of a tourniquet is a ***LAST RESORT*** measure. Use ***ONLY*** when severe, uncontrolled bleeding will cause loss of life. Recognize that long-term use of a tourniquet may cause loss of limb.

(a) Apply tourniquet (TK) band just above bleeding site on limb. A band at least 3 inches (7.5 cm) or wider is best.

(b) Follow steps illustrated in **Figure V-4**.

(c) Use a stick at least 6 inches (15 cm) long.

(d) Tighten only enough to stop arterial bleeding.

(e) Mark a ***TK*** on the forehead with the time applied.

(f) ***DO NOT*** cover the tourniquet.

CAUTION: The following directions apply ***ONLY*** in survival situations where rescue is ***UNLIKELY*** and ***NO*** medical aid is available.

(g) If rescue or medical aid is not available for over 2 hours, an attempt to ***SLOWLY*** loosen the tourniquet may be made 20 minutes after application. Before loosening—

- Ensure pressure dressing is in place.
- Ensure bleeding has stopped
- Loosen tourniquet ***SLOWLY*** to restore circulation.
- Leave loosened tourniquet in position in case bleeding resumes.

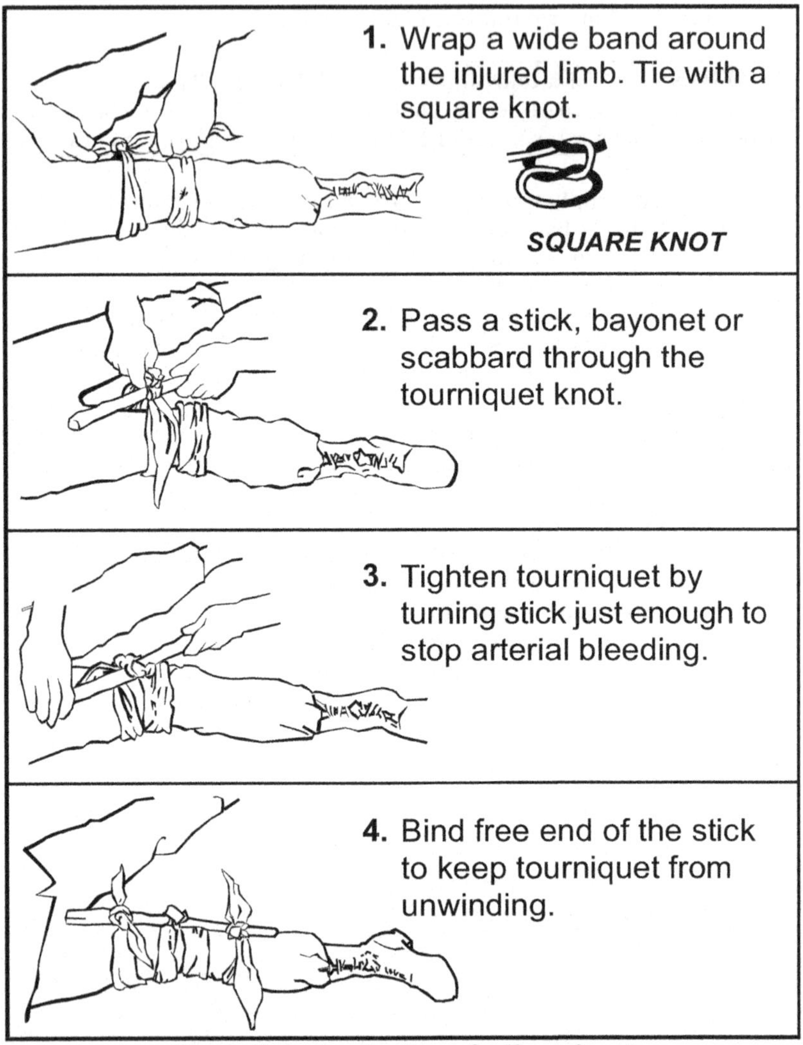

Figure V-4. Application of a Tourniquet

c. Treat shock. (Shock is difficult to identify or treat under field conditions. It may be present with or without visible injury.)

(1) Identify by one or more of the following:

(a) Pale, cool, and sweaty skin.

(b) Fast breathing and a weak, fast pulse.

(c) Anxiety or mental confusion.

(d) Decreased urine output.

(2) Maintain circulation.

(3) Treat underlying injury.

(4) Maintain normal body temperature.

(a) Remove wet clothing.

(b) Give warm fluids.

•***DO NOT*** give fluids to an unconscious victim.

•***DO NOT*** give fluids if they cause victim to gag.

(c) Insulate from ground.

(d) Shelter from the elements.

(5) Place conscious victim on back.

(6) Place very weak or unconscious victim on side, this will—

(a) Allow mouth to drain.

(b) Prevent tongue from blocking airway.

d. Treat chest injuries.

(1) Sucking chest wound. This occurs when chest wall is penetrated; may cause victim to gasp for breath; may cause sucking sound; may create bloody froth as air escapes the chest.

(a) ***Immediately*** seal wound with hand or airtight material.

(b) Tape airtight material over wound on ***3 sides only*** **(Figure V-5)** to allow air to escape from the wound but not to enter.

(c) Monitor breathing and check dressing.

(d) Lift untapped side of dressing as victim ***exhales*** to allow trapped air to escape, as necessary.

(2) Flail chest. Results from blunt trauma when ***3*** or ***more*** ribs are broken in ***2*** or more places. The flail segment is the broken area that moves in a direction opposite to the rest of chest during breathing.

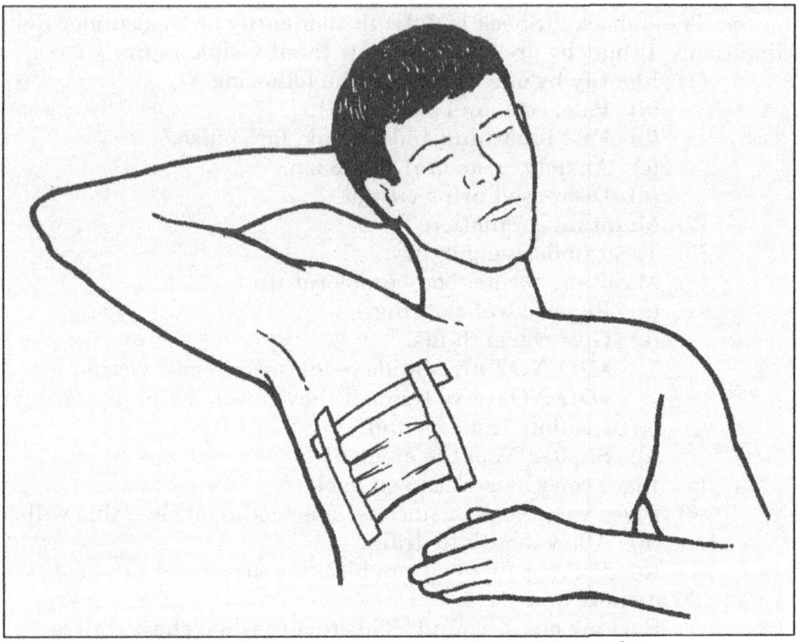

Figure V-5. Sucking Chest Wound Dressing

(a) Stabilize the flail segment as follows:

•Place rolled-up clothing or bulky pad over site.

•Tape pad to site

•***DO NOT*** wrap tape around chest.

(b) Have victim keep segment still with hand pressure.

(c) Roll victim onto side of flail segment injury (as other injuries allow).

(3) Fractured ribs.

(a) Encourage deep breathing (painful, but necessary to prevent the possible development of pneumonia).

(b) ***DO NOT*** constrict breathing by taping ribs.

e. Treat fractures, sprains, and dislocations.

(1) Control bleeding.

(2) Remove watches, jewelry, and constrictive clothing.

(3) If fracture penetrates the skin—

(a) Clean wound by gentle irrigation with water.
(b) Apply dressing over wound.
(4) Position limb as normally as possible.
(5) Splint in position found (if ***unable*** to straighten limb).
(6) Improvise a splint with available materials:
(a) Sticks or straight, stiff materials from equipment.
(b) Body parts (for example, opposite leg, arm-to-chest).
(7) Attach with strips of cloth, parachute cord, etc.
(8) Keep the fractured bones from moving by immobilizing the joints on both sides of the fracture. If fracture is in a joint, immobilize the bones on both sides of the joint.

CAUTION: Splint fingers in a slightly flexed position, ***NOT*** in straight position. Hand should look like it is grasping an apple.

(9) Use ***RICES*** treatment for 72 hours.
(a) **R**est.
(b) **I**ce.
(c) **C**ompression.
(d) **E**levation.
(e) **S**tabilization.
(10) Apply cold to acute injuries.
(11) Use 15 to 20 minute periods of cold application.
(a) ***DO NOT*** use continuous cold therapy.
(b) Repeat 3 to 4 times per day.
(c) Avoid cooling that can cause frostbite or hypothermia.
(12) Wrap with a compression bandage after cold therapy.
(13) Elevate injured area above heart level to reduce swelling.
(14) Check periodically for a pulse beyond the injury site.
(15) Loosen bandage or reapply splint if no pulse is felt or if swelling occurs because bandage is too tight.

2. Common Injuries and Illnesses

a. Burns.
(1) Cool the burned area with water.
(a) Use immersion or cool compresses.
(b) Avoid aggressive cooling with ice or frigid water.
(2) Remove watches, jewelry, constrictive clothing.

(3) ***DO NOT*** remove embedded, charred material that will cause burned areas to bleed.

(4) Cover with sterile dressings.

(5) ***DO NOT*** use lotion or grease.

(6) Avoid moving or rubbing the burned part.

(7) Drink ***extra*** water to compensate for increased fluid loss from burns. (Add ***1/4 teaspoon*** of ***salt*** [if available] to ***each quart*** of ***water***.)

(8) Change dressings when soaked or dirty.

b. Eye injuries.

(1) Sun/snow blindness (gritty, burning sensation, and possible reduction in vision caused by sun exposure).

(a) Prevent with improvised goggles. **(See Chapter VI, page VI-3, Figure VI-2.)**

(b) Treat by patching affected eye(s).

•Check after 12 hours.

•Replace patch for another 12 hours if not healed.

(c) Use cool compresses to reduce pain.

(2) Foreign body in eye.

(a) Irrigate with clean water from the ***inside*** to the ***outside*** corner of the eye.

(b) If foreign body is not removed by irrigation, improvise a small swab. Moisten and wipe gently over the affected area.

(c) If foreign body is ***STILL*** not removed, patch eye for 24 hours and then reattempt removal using steps **(a)** and **(b)**.

c. Heat injury.

(1) Heat cramps (cramps in legs or abdomen).

(a) Rest.

(b) Drink water. Add ***1/4 teaspoon*** of salt ***per quart***.

(2) Heat exhaustion (pale, sweating, moist, cool skin).

(a) Rest in shade.

(b) Drink water.

(c) Protect from further heat exposure.

(3) Heat stroke (victim disoriented or unconscious, skin is hot and flushed [sweating **may** or **may not** occur], fast pulse).

CAUTION: Handle heat stroke victim gently. Shock, seizures, and cardiac arrest can occur.

(a) Cool as rapidly as possible (saturate clothing with water and fan the victim). Remember to cool the groin and armpit areas. (Avoid overcooling.)

(b) Maintain airway, breathing, and circulation.

d. Cold injuries:

(1) Frostnip and frostbite—

(a) Are progressive injuries.

•Ears, nose, fingers, and toes are affected first.

•Areas will feel cold and may tingle leading to—

••Numbness that progresses to—

•••Waxy appearance with stiff skin that cannot glide freely over a joint.

(b) Frostnipped areas rewarm with body heat. If body heat ***WILL NOT*** rewarm area in 15 to 20 minutes, then frostbite is present.

(c) Frostbitten areas are deeply frozen and require medical treatment.

CAUTION: In frostbite, repeated freezing and thawing causes severe pain and increases damage to the tissue. ***DO NOT*** rub frozen tissue. ***DO NOT*** thaw frozen tissue.

(2) Hypothermia—

(a) Is a progressive injury.

•Intense shivering with impaired ability to perform complex tasks leads to—

••Violent shivering, difficulty speaking, sluggish thinking go to—

•••Muscular rigidity with blue, puffy skin; jerky movements go to—

••••Coma, respiratory and cardiac failure.

(b) Protect victim from the environment as follows:

•Remove wet clothing.

•Put on dry clothing (if available).

•Prevent further heat loss.

••Cover top of head.

••Insulate from above and below.

•Warm with blankets, sleeping bags, or shelter.

•Warm central areas before extremities.

••Place heat packs in groin, armpits, and around neck.

••Avoid causing burns to skin.

CAUTION: Handle hypothermia victim gently. Avoid overly rapid rewarming which may cause cardiac arrest. Rewarming of victim with skin-to-skin contact by volunteer(s) inside of a sleeping bag is a survival technique but can cause internal temperatures of all to drop.

e. Skin tissue damage.

(1) Immersion injuries. Skin becomes wrinkled as in ***dishpan hands***.

(a) Avoid walking on affected feet.

(b) Pat dry; ***DO NOT*** rub. Skin tissue will be sensitive.

(c) Dry socks and shoes. Keep feet protected.

(d) Loosen boots, cuffs, etc., to improve circulation.

(e) Keep area dry, warm, and open to air.

(f) ***DO NOT*** apply creams or ointments.

(2) Saltwater sores.

(a) Change body positions frequently.

(b) Keep sores dry.

(c) Use antiseptic (if available).

(d) ***DO NOT*** open or squeeze sores.

f. Snakebite.

CAUTION: This snakebite treatment recommendation is for situations where medical aid and specialized equipment are not available.

(1) Nonpoisonous. Clean and bandage wound.

(2) Poisonous.

(a) Remove constricting items.

(b) Minimize activity.

(c) ***DO NOT*** cut the bite site; ***DO NOT*** use your mouth to create suction.

(d) Clean bite with soap and water; cover with a dressing.

(e) Overwrap the bite site with a tight (elastic) bandage **(Figure V-6)**. The intent is to slow capillary and venous blood flow but not arterial flow. Check for pulse below the overwrap.

(f) Splint bitten extremity to prevent motion.

(g) Treat for shock **(page V-7, paragraph 1c)**.

(h) Position extremity below level of heart.

(i) Construct shelter if necessary (let the victim rest).

(j) For conscious victims, force fluids.

g. Marine life.

(1) Stings.

(a) Flush wound with salt water (fresh water stimulates toxin release).

(b) Remove jewelry and watches.

(c) Remove tentacles and gently scrape or shave skin.

(d) Apply a steroid cream (if available).

(e) ***DO NOT*** rub area with sand.

(f) Treat for shock; artificial respiration may be required **(page V-1, paragraph 1a)**.

(g) ***DO NOT*** use urine to flush or treat wounds.

(2) Punctures.

(a) Immerse affected part in hot water or apply hot compresses for 30-60 minutes (as hot as victim can tolerate).

(b) Cover with clean dressing.

(c) Treat for shock as needed.

h. Skin irritants (includes poison oak and poison ivy).

(1) Wash with large amounts of water. Use soap (if available).

(2) Keep covered to prevent scratching.

i. Infection.

(1) Keep wound clean.

(2) Use iodine tablet solution or diluted betadine to prevent or treat infection.

(3) Change bandages as needed.

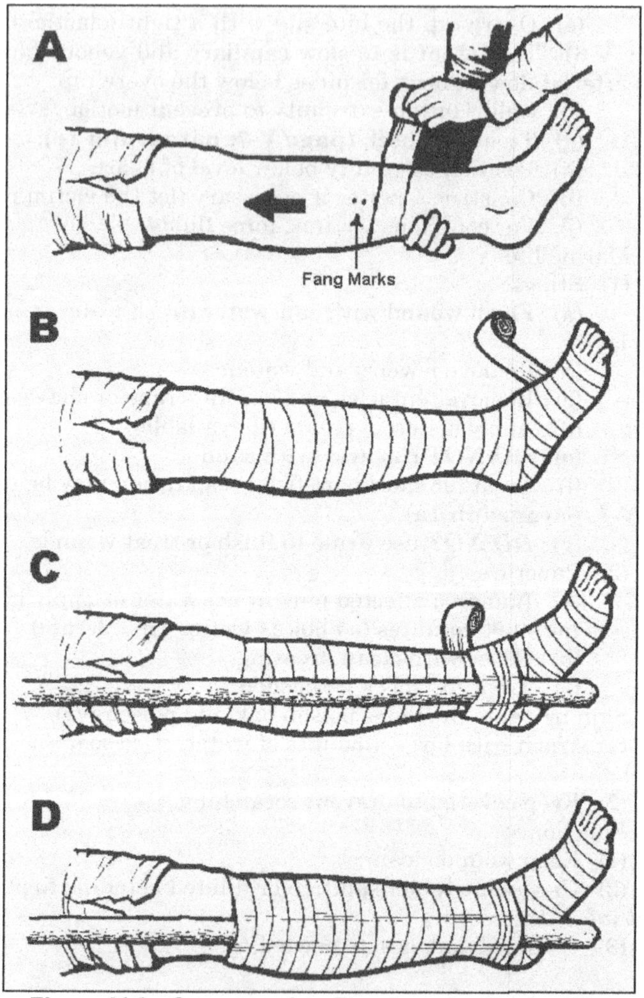

Figure V-6. Compression Bandage for Snake Bite

j. Dysentery and diarrhea.

(1) Drink ***extra*** water.

(2) Use a liquid diet.

(3) Eat charcoal. Make a paste by mixing fine charcoal particles with water. (It may relieve symptoms by absorbing toxins.)

k. Constipation (can be expected in survival situations).

(1) ***DO NOT*** take laxatives.

(2) Exercise.

(3) Drink **extra** water.

3. Plant Medicine

a. Tannin.

(1) Medical uses. Burns, diarrhea, dysentery, skin problems, and parasites. Tannin solution prevents infection and aids healing.

(2) Sources. Found in the outer bark of all trees, acorns, banana plants, common plantain, strawberry leaves, and blackberry stems.

(3) Preparation.

(a) Place crushed outer bark, acorns, or leaves in water.

(b) Leach out the tannin by soaking or boiling.

- Increase tannin content by longer soaking time.
- Replace depleted material with fresh bark/plants.

(4) Treatments.

(a) Burns.

- Moisten bandage with cooled tannin tea.
- Apply compress to burned area.
- Pour cooled tea on burned areas to ease pain.

(b) Diarrhea, dysentery, and worms. Drink strong tea solution (may promote voiding of worms).

(c) Skin problems (dry rashes and fungal infections). Apply cool compresses or soak affected part to relieve itching and promote healing.

(d) Lice and insect bites. Wash affected areas with tea to ease itching.

b. Salicin/salicylic acid.

(1) Medical uses. Aches, colds, fever, inflammation, pain, sprains, and sore throat (aspirin-like qualities).

(2) Sources. Willow and aspen trees **(Figure V-7)**.

(3) Preparation.

(a) Gather twigs, buds, or cambium layer (soft, moist layer between the outer bark and the wood) of willow or aspen.

(b) Prepare tea as described in paragraph **3a(3)**.

(c) Make poultice.

•Crush the plant or stems.

•Make a pulpy mass.

(4) Treatments.

(a) Chew on twigs, buds, or cambium for symptom relief.

(b) Drink tea for colds and sore throat.

(c) Use warm, moist poultice for aches and sprains.

•Apply pulpy mass over injury.

•Hold in place with a dressing.

c. Common plantain.

(1) Medical uses. Itching, wounds, abrasions, stings, diarrhea, and dysentery.

(2) Source. There are over 200 plantain species with similar medicinal properties. The common plantain is shown in **Figure V-7**.

(3) Preparation.

(a) Brew tea from seeds.

(b) Brew tea from leaves.

(c) Make poultice of leaves.

(4) Treatments.

(a) Drink tea made from seeds for diarrhea or dysentery.

(b) Drink tea made from leaves for vitamin and minerals.

(c) Use poultice to treat cuts, sores, burns, and stings.

d. Papain.

(1) Medical uses. Digestive aid, meat tenderizer, and a food source.

(2) Source. Fruit of the papaya tree **(Figure V-7)**.

(3) Preparation.

(a) Make cuts in **unripe** fruit.

(b) Gather milky white sap for its papain content.

(c) Avoid getting sap in eyes or wounds.

(4) Treatments.

(a) Use sap to tenderize tough meat.

(b) Eat **ripe** fruit for food, vitamins, and minerals.

e. Common Cattail.

(1) Medical uses. Wounds, sores, boils, inflammations, burns, and an excellent food source.

(2) Source. Cattail plant found in marshes **(Figure V-7)**.

(3) Preparation.

(a) Pound roots into a pulpy mass for a poultice.

(b) Cook and eat green bloom spikes.

(c) Collect yellow pollen for flour substitute.

(d) Peel and eat tender shoots (raw or cooked).

(4) Treatments.

(a) Apply poultice to affected area.

(b) Use plant for food, vitamins, and minerals.

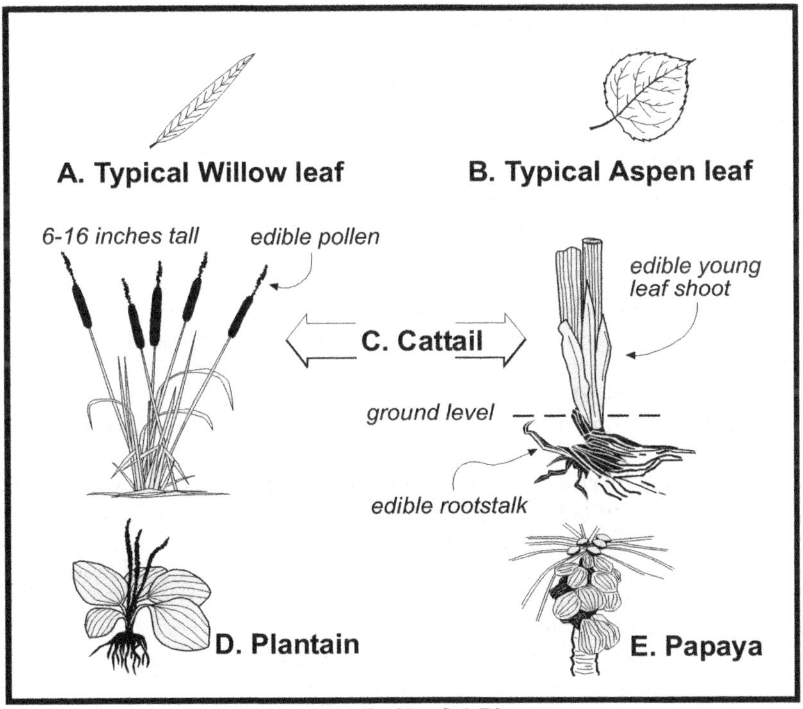

Figure V-7. Useful Plants

4. Health and Hygiene

a. Stay clean (daily regimen).

(1) Minimize infection by washing. (Use white ashes, sand, or loamy soil as soap substitutes.)

(2) Comb and clean debris from hair.

(3) Cleanse mouth and brush teeth.

(a) Use hardwood twig as toothbrush (fray it by chewing on one end then use as brush).

(b) Use single strand of an inner core string from parachute cord for dental floss.

(c) Use clean finger to stimulate gum tissues by rubbing.

(d) Gargle with salt water to help prevent sore throat and aid in cleaning teeth and gums.

(4) Clean and protect feet.

(a) Change and wash socks

(b) Wash, dry, and massage.

(c) Check frequently for blisters and red areas.

(d) Use adhesive tape/mole skin to prevent damage.

b. Exercise daily.

c. Prevent and control parasites.

(1) Check body for lice, fleas, ticks, etc.

(a) Check body regularly.

(b) Pick off insects and eggs (***DO NOT*** crush).

(2) Wash clothing and use repellents.

(3) Use smoke to fumigate clothing and equipment.

5. Rules for Avoiding Illness

a. Purify all water obtained from natural sources by using iodine tablets, bleach, or boiling for 5 minutes.

b. Locate latrines 200 feet from water and away from shelter.

c. Wash hands before preparing food or water.

d. Clean all eating utensils after each meal.

e. Prevent insect bites by using repellent, netting, and clothing.

f. Dry wet clothing as soon as possible.

g. Eat varied diet.

h. Try to get 7-8 hours sleep per day.

Chapter VI

PERSONAL PROTECTION

1. Priorities

a. Evaluate available resources and situation, then accomplish individual tasks accordingly.

b. First 24 hours in order of situational needs—

(1) Construct survival shelter according to selection criteria.

(2) Procure water.

(3) Establish multiple survival signals.

(4) Build Fire.

c. Second 24 hours in order of situational needs—

(1) Construct necessary tools and weapons.

(2) Procure food.

2. Care and Use of Clothing

a. Never discard clothing.

b. Wear loose and layered clothing.

(1) Tight clothing restricts blood flow regulating body temperature.

(2) Layers create more dead air space.

c. Keep entire body covered to prevent sunburn and dehydration in hot climates. When fully clothed, the majority of body heat escapes through the head and neck areas.

d. Avoid overheating.

(1) Remove layers of clothing before strenuous activities.

(2) Use a hat to regulate body heat.

(3) Wear a hat when in direct sunlight (in hot environment).

e. Dampen clothing when on the ocean in hot weather.

(1) Use salt water, ***NOT*** drinking water.

(2) Dry clothing before dark to prevent hypothermia.

f. Keep clothing dry to maintain its insulation qualities (dry damp clothing in the sun or by a fire).

g. If you fall into the water in the winter—

(1) Build fire.

(2) Remove wet clothing and rewarm by fire.

(3) Finish drying clothing by fire.

h. If no fire is available—

(1) Remove clothing and get into sleeping bag (if available).

(2) Allow wet clothes to freeze.

(3) Break ice out of clothing.

i. Keep clothing clean (dirt reduces its insulation qualities). Examine clothing frequently for damage.

(1) ***DO NOT*** sit or lie directly on the ground.

(2) Wash clothing whenever possible.

(3) Repair when necessary by using—

(a) Needle and thread.

(b) Safety pins.

(c) Tape.

j. Improvised foot protection **(Figure VI-1)**.

(1) Cut **2** to **4** layers of cloth into a 30-inch square.

(2) Fold into a triangle.

(3) Center foot on triangle with toes toward corner.

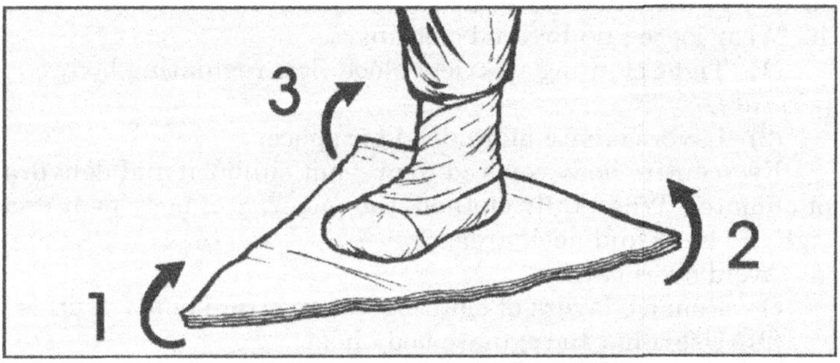

Figure VI-1 Improvised Foot Wear

(4) Fold front over the toes.

(5) Fold side corners, one at a time, over the instep.

(6) Secure by rope, vines, tape, etc., or tuck into other layers of material.

3. Other Protective Equipment

a. Sleeping bag.

(1) Fluff before use, ***especially*** at foot of bag.

(2) Air and dry daily to remove body moisture.

(3) Improvise with available material, dry grass, leaves, dry moss, etc.

b. Sun and snow goggles **(Figure VI-2)**.

(1) Wear in bright sun or snow conditions.

(2) Improvise by cutting small horizontal slits in webbing, bark, or similar materials.

Figure VI-2. Sun and Snow Goggles

c. Gaiters **(Figure VI-3)**. Used to protect from sand, snow, insects, and scratches (wrap material around lower leg and top of boots).

Figure VI-3. Gaiters

4. Shelters

Evasion considerations apply.

a. Site selection.

(1) Near signal and recovery site.

(2) Available food and water.

(3) Avoid natural hazards:

(a) Dead standing trees.

(b) Drainage and dry river beds except in combat areas.

(c) Avalanche areas.

(4) Location large and level enough to lie down in.

b. Types.

(1) Immediate shelters. Find shelter needing minimal improvements **(Figure VI-4)**.

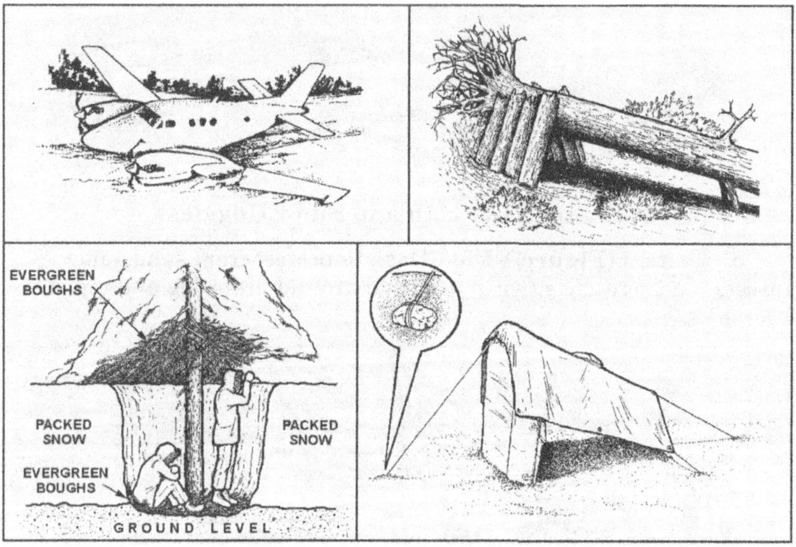

Figure VI-4. Immediate Shelters

(2) General shelter. Temperate climates require any shelter that gives protection from wind and rain.

(3) Thermal A Frame, Snow Trench, Snow Cave. **(Figures VI-5 through VI-7)**. Cold climates require an enclosed, insulated shelter.

(a) Snow is the most abundant insulating material.

(b) Air vent is required to prevent carbon monoxide poisoning when using an open flame inside enclosed shelters.

Note: As a general rule, unless you can see your breath, your snow shelter is too warm and should be cooled down to preclude melting and dripping.

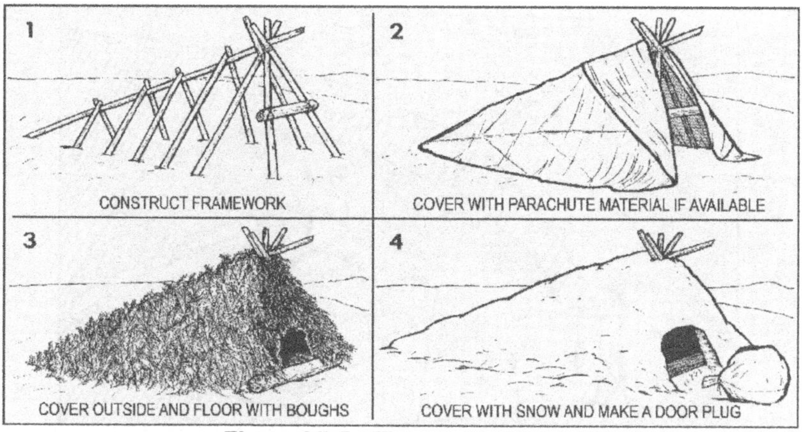

Figure VI-5. Thermal A Frame

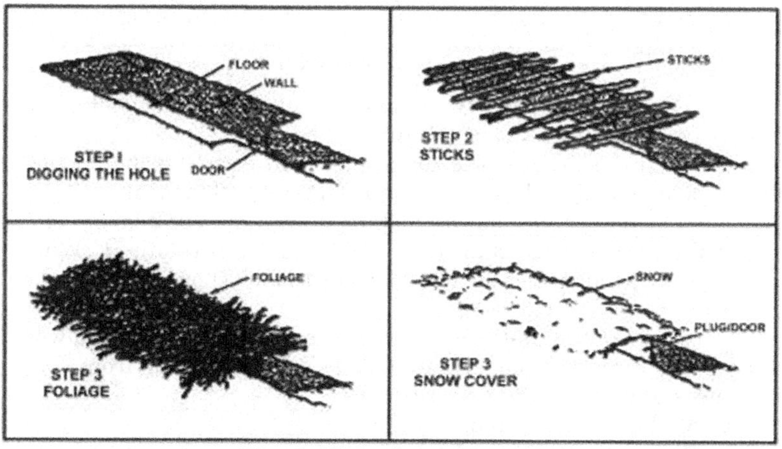

Figure VI-6. Snow Trench

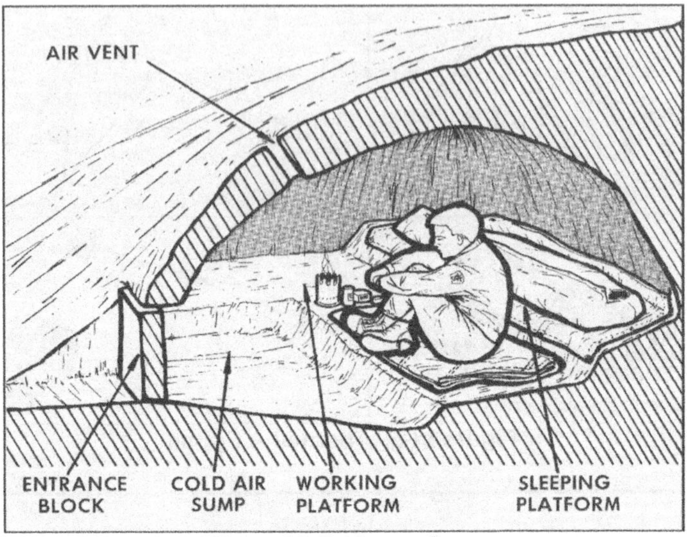

Figure VI-7. Snow Cave

(4) Shade shelter. Hot climates require a shade shelter to protect from ultraviolet rays **(Figure VI-8)**.

(a) To reduce the surface temperature, the shelter floor should be elevated or dug down (approximately 18 inches).

(b) For thermal protection, a minimum of **2** layers of material suspended 12-18 inches above the head is required. White is the best color to reflect heat (inner most layer should be of darker material).

(5) Elevated platform shelter **(Figure VI-9)**. Tropical/wet climates require enclosed, elevated shelter for protection from dampness and insects.

c. Shelter construction.

(1) Have entrance 45-90 degrees from prevailing wind.

(2) Cover with available material.

(a) If natural materials are used, arrange them in layers starting at the bottom with each layer overlapping the previous one. See **Figure VI-10** for an example.

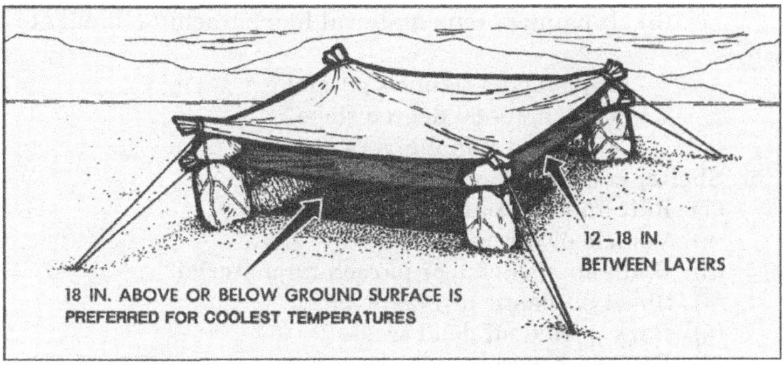

Figure VI-8. Poncho/Parachute Shade Shelter

Figure VI-9. Elevated Platform Shelter

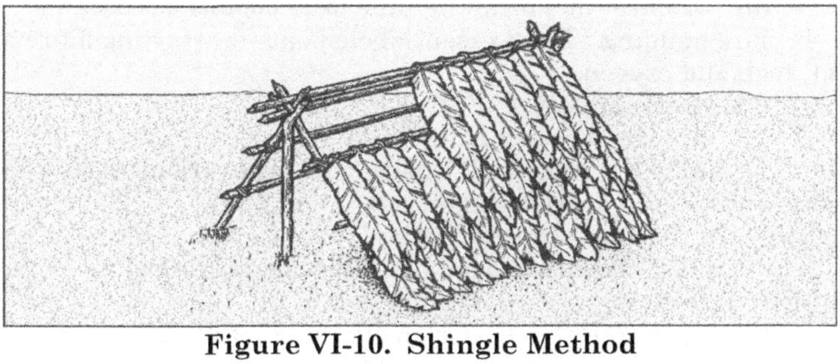
Figure VI-10. Shingle Method

(b) If using porous material like parachute, blankets, etc.—

- Stretch as tight as possible
- Use a 40–60 degree slope.
- Use additional layers in heavy rain.

d. Shelter construction materials:

(1) Raft and raft parts.

(2) Vehicle or aircraft parts.

(3) Blankets, poncho, or parachute material.

(4) Sheet of plastic or plastic bag.

(5) Bark peeled off dead trees.

(6) Boughs, broad leaves, dry moss.

(7) Grass and sod.

(8) Snow.

(9) Sand and rocks.

e. Bed construction. Construct a bed to protect from cold, damp, ground using—

(1) Raft or foam rubber from vehicle seats.

(2) Boughs, leaves, or dry moss.

5. Fires

CAUTION: Weigh hazards and risks of detection against the need for a fire.

a. Evasion considerations:

(1) Use trees or other sources to dissipate smoke.

(2) Use fires at dusk, dawn, or during inclement weather.

(3) Use fires at times when the local populace is cooking.

b. Fire building. The **3** essential elements for starting a fire are heat, fuel, and oxygen.

(1) Heat sources:

(a) Matches or lighter.

(b) Flint and steel (experiment with various rocks and metals until a good spark is produced).

(c) Sparks from batteries.

(d) Concentrated sunlight (use magnifying glass or flashlight reflectors).

(e) Pyrotechnics, such as flares **(last resort)**, etc.

(f) Friction method **(Figure VI-11)**. Without prior training, this method is difficult to master and requires a lot of time to build the device.

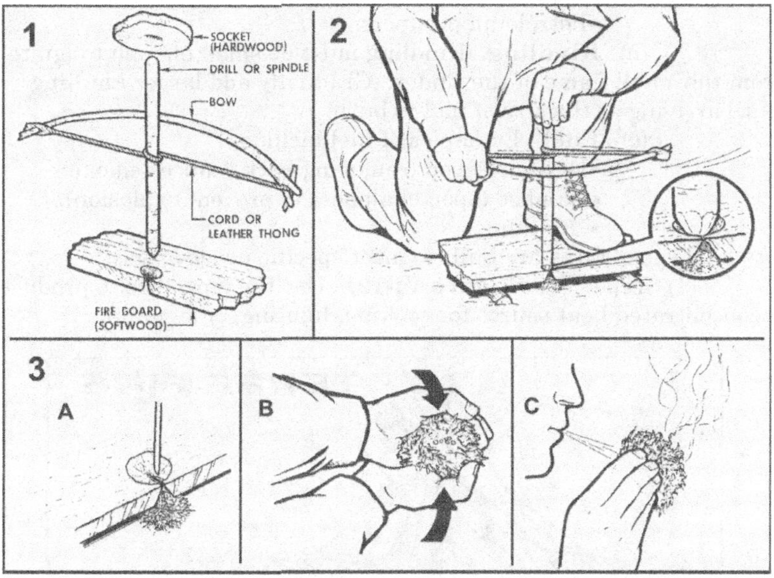

Figure VI-11. Friction Method

> **Note:** If possible, carry a fire-starting device with you.

(2) Fuel is divided into **3** categories: tinder, kindling, and fuel. (Gather large amounts of each category before igniting the fire.)

(a) **Tinder**. Tinder must be very finely shaved or shredded to provide a low combustion point and fluffed to allow oxygen to flow through. (To get tinder to burn hotter and longer, saturate with Vaseline, Chapstick, insect repellant, aircraft fuel, etc.) Examples of tinder include—

- Cotton.
- Candle (shred the wick, not the wax).
- Plastic spoon, fork, or knife.
- Foam rubber.

•Dry bark.
•Dry grasses.
•Gun powder.
•Pitch.
•Petroleum products.

(b) **Kindling**. Kindling must be small enough to ignite from the small flame of the tinder. Gradually add larger kindling until arriving at the size of fuel to burn.

(c) **Fuel**. Examples of fuel include—

•Dry hardwood (removing bark reduces smoke).
•Bamboo (open chambers to prevent explosion).
•Dry dung.

c. Types. Fires are built to meet specific needs or uses.

(1) Tepee fire **(Figure VI-12)**. Use the tepee fire to produce a concentrated heat source for cooking, lighting, or signaling.

Figure VI-12. Tepee Fire

(2) Log cabin fire **(Figure VI-13)**. Use the log cabin fire to produce large amounts of light and heat, to dry out wet wood, and provide coals for cooking, etc.

Figure VI-13. Log Cabin or Pyramid Fires

(3) Sod fire and reflector (**Figure VI-14**). Use fire reflectors to get the most warmth from a fire. Build fires against rocks or logs.

CAUTION: ***DO NOT*** use porous rocks or riverbed rock—they may explode when heated.

Figure VI-14. Sod Fire and Reflector

(4) Dakota fire hole (Figure VI-15). Use the Dakota fire hole for high winds or evasion situations.

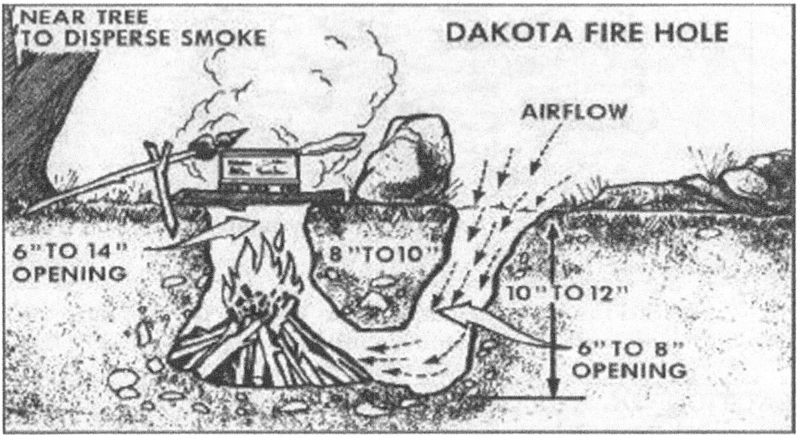

Figure VI-15. Dakota Fire Hole

(5) Improvised stoves **(Figure VI-16)**. These are very efficient.

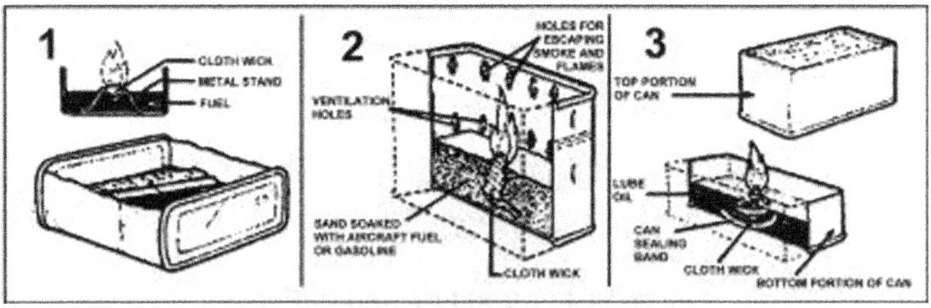

Figure VI-16. Improvised Stove

Chapter VII

Water

1. Water Requirements

Drink **extra** water. Minimum 2 quarts per day to maintain fluid level. Exertion, heat, injury, or an illness increases water loss.

> **Note:** Pale yellow urine indicates adequate hydration.

2. Water Procurement

a. ***DO NOT*** drink—

(1) Urine.
(2) Fish juices.
(3) Blood.
(4) Sea water.
(5) Alcohol.
(6) Melted water from new sea ice.

b. Water sources:

(1) Surface water (streams, lakes, and springs).
(2) Precipitation (rain, snow, dew, sleet) **(FigureVII-1)**.
(3) Subsurface (wells and cisterns).
(4) Ground water (when no surface water is available) **(Figure VII-2)**.

(a) Abundance of lush green vegetation.
(b) Drainages and low-lying areas.
(c) *"V"* intersecting game trails often point to water.
(d) Presence of swarming insects indicates water is near.
(e) Bird flight in the early morning or late afternoon might indicate the direction to water.

(5) Snow or ice.

(a) ***DO NOT*** eat ice or snow.

- Lowers body temperature.
- Induces dehydration.
- Causes minor cold injury to lips and mouth.

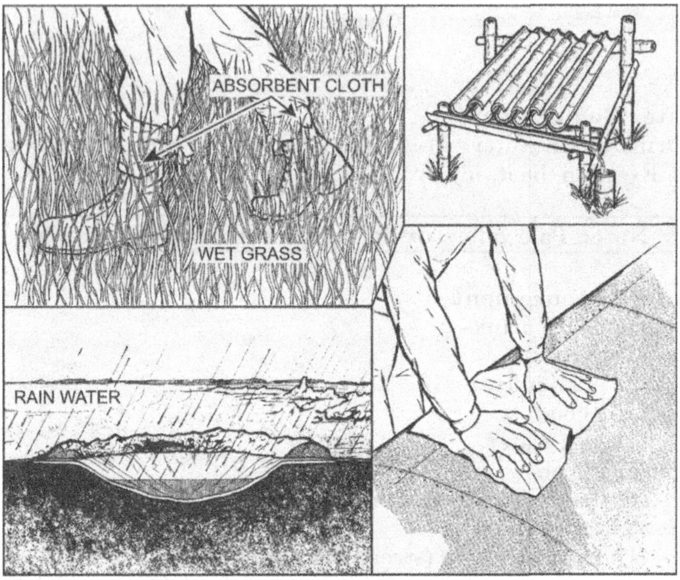

Figure VII-1. Water Procurement

Figure VII-2. Water Indicators

(b) Melt with fire.
- Stir frequently to prevent damaging container.
- Speed the process by adding hot rocks or water.

(c) Melt with body heat.
- Use waterproof container.
- Place between layers of clothing.
- ***DO NOT*** **place next to the skin.**

(d) Use a water generator **(Figure VII-3)**.

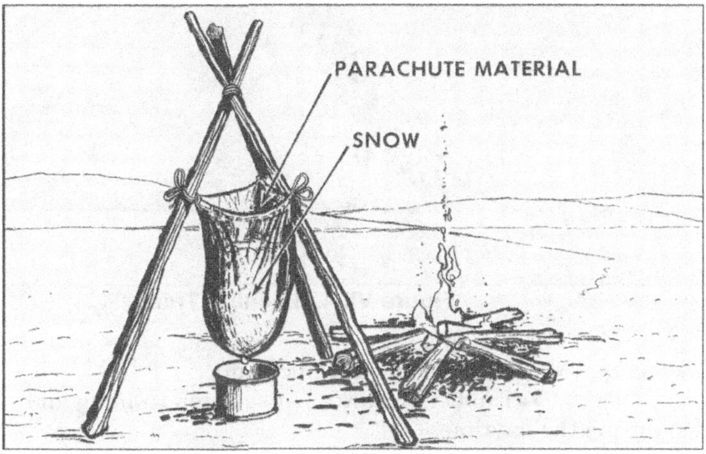

Figure VII-3. Water Generator

(6) Open seas.

(a) Water available in survival kits.

(b) Precipitation.
- Drink as much as possible.
- Catch rain in spray shields and life raft covers.
- Collect dew off raft.

(c) Old sea ice or icebergs **(Table VII-1)**.

Table VII-1. Old Sea Ice or Icebergs

OLD SEA ICE	NEW SEA ICE
Bluish or blackish	Milky or grey
Shatters easily	Does not break easily
Rounded corners	Sharp edges
Tastes relatively salt-free	Tastes extremely salty

(7) Tropical areas.

(a) All open sources previously mentioned.

(b) Vegetation.

•Plants with hollow sections can collect moisture.

•Leaning Tree. Cloth absorbs rain running down tree and drips into container **(Figure VII-4)**.

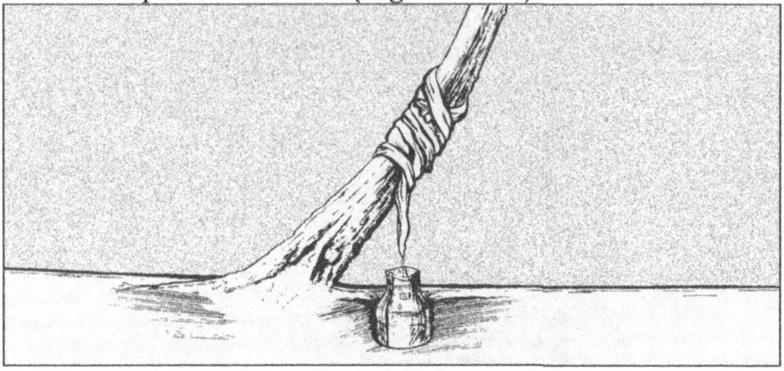

Figure VII-4. Leaning Tree

•Banana plants.

•Water trees (avoid milky sap).

••Tap before dark. Let sap stop running and harden during the daytime.

••Produce most water at night.

••For evasion situations, bore into the roots and collect water.

•Vines **(Figure VII-5A)**.

••Cut bark (***DO NOT*** use milky sap).

••If juice is clear and water like, cut as large a piece of vine as possible (cut the top first).

••Pour into hand to check smell, color, and taste to determine if drinkable.

••***DO NOT*** touch vine to lips.

••When water flow stops, cut off 6 inches of opposite end, water will flow again.

•Old bamboo.

••Shake and listen for water.

••Bore hole at bottom of section to obtain water.

••Cut out entire section to carry with you.

••Filter and purify.

•Green bamboo **(Figure VII-5B)**.

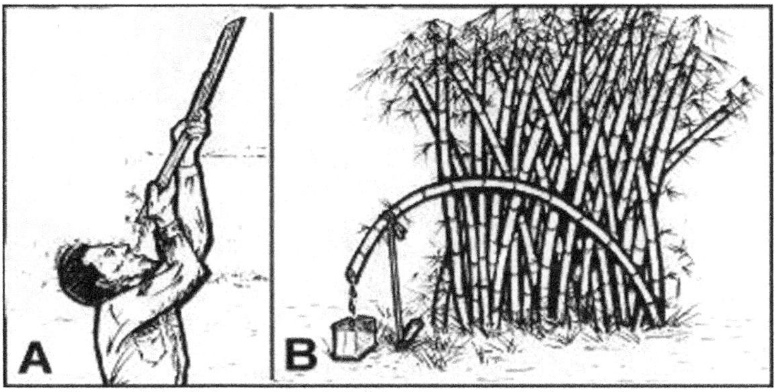

Figure VII-5 A and B. Water Vines and Green Bamboo

> **CAUTION:** Liquid contained in green coconuts (**ripe** coconuts may cause diarrhea).

•Beach well. Along the coast, obtain water by digging a beach well **(Figure VII-6)**.

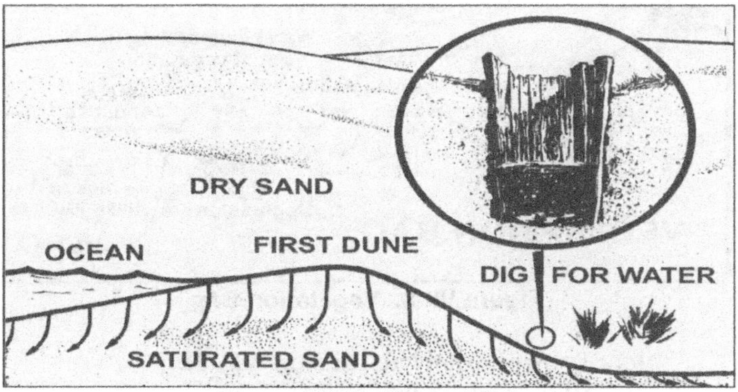

Figure VII-6. Beach Well

(8) Dry areas.

(a) Solar still **(Figure VII-7)**.

(b) Vegetation bag **(Figure VII-8)**.

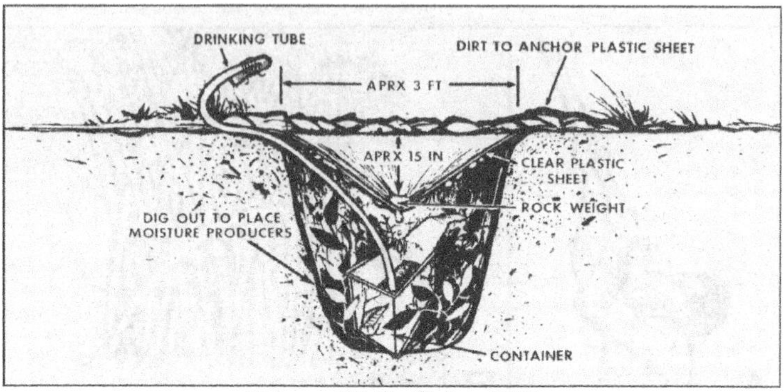

Figure VII-7. Solar Still

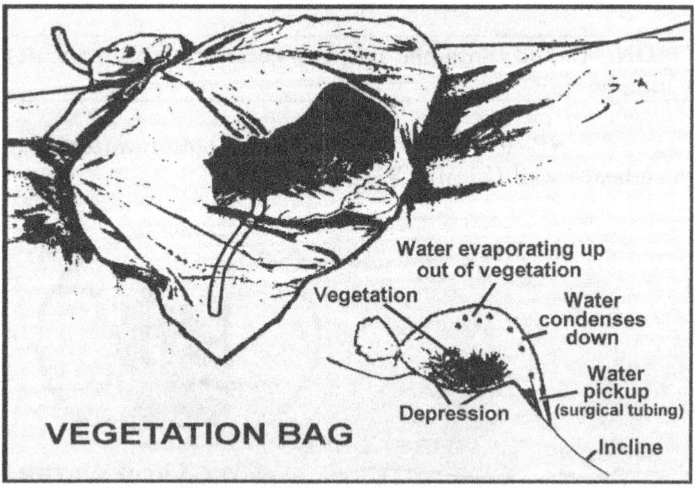

Figure VII-8. Vegetation Bag

(c) Transpiration bag **(Figure VII-9)**.

•Water bag must be clear.

•Water will taste like the plant smells.

(d) Seepage basin **(Figure VII-10)**.

CAUTION: ***DO NOT*** use poisonous/toxic plants in vegetation/ transpiration bags.

Figure VII-9. Transpiration Bag

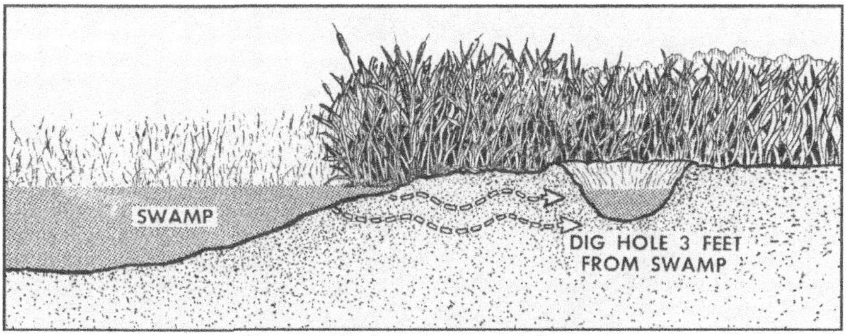

Figure VII-10. Seepage Basin

3. Water Preparation and Storage

a. Filtration. Filter through porous material (sand/charcoal).

b. Purification.

(1) Water from live plants requires no further treatment.

(2) Purify all other water.

(a) Boil at least 1 minute.

(b) Pour from one container to another to improve taste to aerate.

(c) Water purification tablets. Follow instructions on package.

c. Potable Water.

(1) If water cannot be purified, obtain water from a clear, cold, clean, and fast running source (if possible).

(2) Put in clear container and expose to the sun's ultraviolet rays to kill bacteria.

d. Storage. To prevent contamination, use a clean, covered or sealed container.

(1) Trash bag.

(2) Prophylactic.

(3) Section of bamboo.

(4) Flotation gear.

Chapter VIII
FOOD

1. Food Procurement

a. Sources and location.

(1) Mammals can be found where—

(a) Trails lead to watering, feeding, and bedding areas.

(b) Droppings or tracks look fresh.

(2) Birds can be found by—

(a) Observing the direction of flight in the early morning and late afternoon (leads to feeding, watering, and roosting areas).

(b) Listening for bird noises (indication of nesting areas

(3) Fish and other marine life locations **(Figure VIII-1)**.

Figure VIII-1. Fishing Locations

(4) Reptiles and amphibians are found almost worldwide.

(5) Insects are found—

(a) In dead logs and stumps.

(b) At ant and termite mounds.

(c) On ponds, lakes, and slow moving streams.

b. Procurement techniques.

(1) Snares—

(a) Work while unattended.

(b) Location:
•Trails leading to water, feeding, and bedding areas.
•Mouth of dens **(Figure VIII-2)**.

Figure VIII-2. Snare Placement

(c) Construction of simple loop snare.
•Use materials that will not break under the strain of holding an animal.
•Use a figure 8 (locking loop) if wire is used **(Figure VIII-3)**.
••Once tightened, the wire locks in place, preventing reopening, and the animal's escape.
•To construct a squirrel pole **(Figure VIII-4)** use simple loop snares.
•Make noose opening slightly larger than the animal's head **(3-finger** width for squirrels, **fist-sized** for rabbits**)**.

(d) Placement of snares (set as many as possible).
•Avoid disturbing the area.
•Use funneling (natural or improvised) **(Figure VIII-5)**.

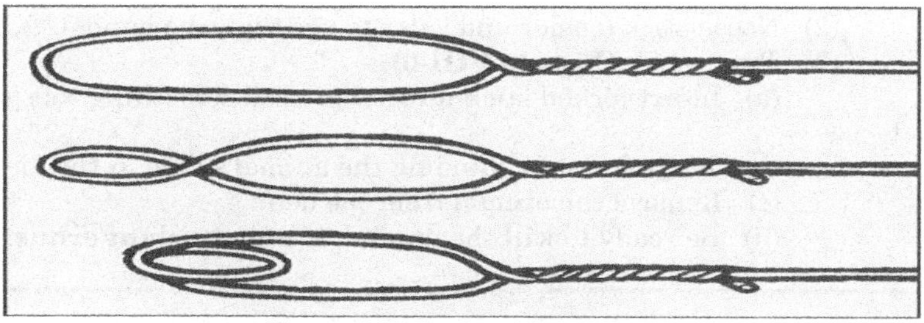

Figure VIII-3. Locking Loop

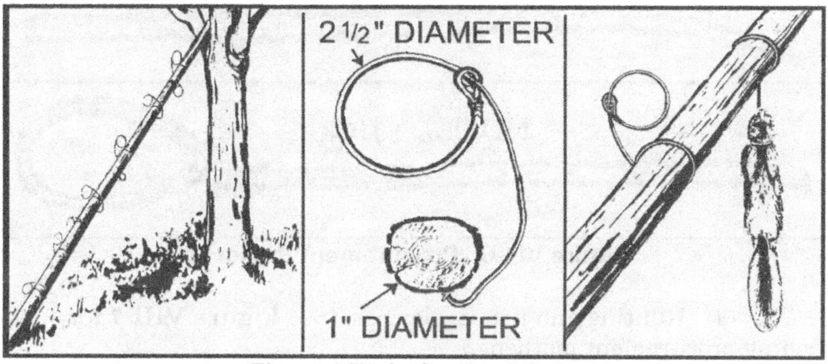

Figure VIII-4. Squirrel Pole

Figure VIII-5. Funneling

(2) Noose stick (easier and safer to use than the hands).

(3) Twist stick **(Figure VIII-6)**.

(a) Insert forked stick into a den until something soft is met.

(b) Twist the stick, binding the animal's hide in the fork.

(c) Remove the animal from the den.

(d) Be ready to **kill** the animal; **it may be dangerous**.

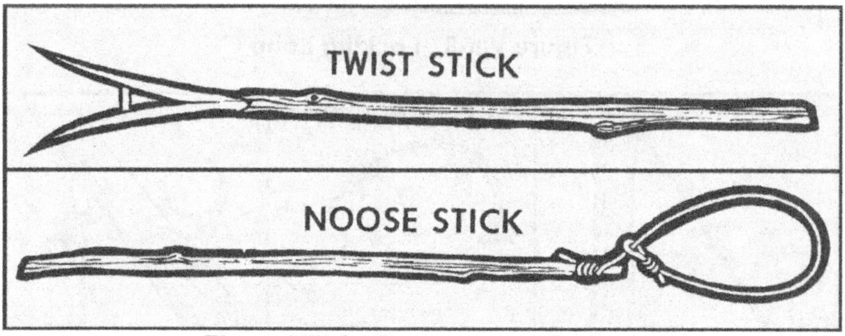

Figure VIII-6. Procurement Devices

(4) Hunting and fishing devices. (See **Figure VIII-7** for fishing procurement methods.)

(a) Club or rock.

(b) Spear.

(c) Slingshot.

(d) Pole, line, and hook.

(e) Net.

(f) Trap.

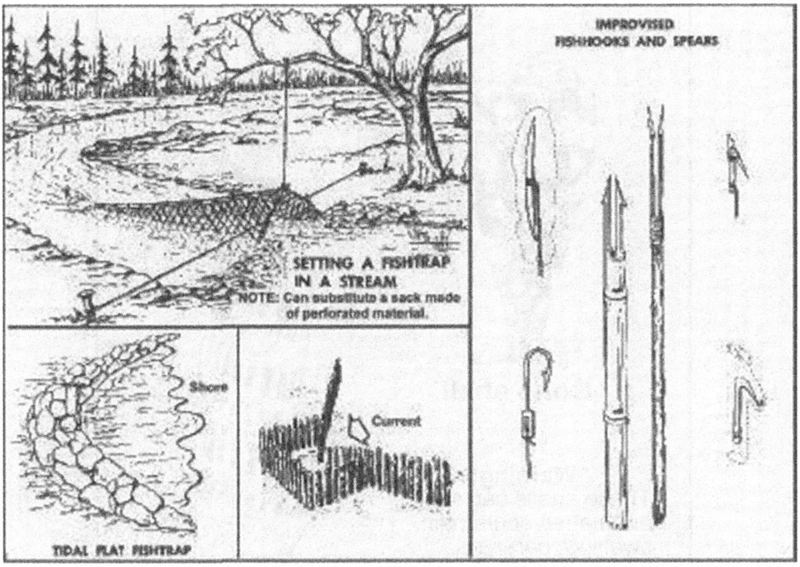

Figure VIII-7. Procurement Methods

(5) Precautions:

(a) Wear shoes to protect the feet when wading in water.

(b) Avoid reaching into dark holes.

(c) **Kill** animals before handling. Animals in distress may attract the enemy.

(d) ***DO NOT*** secure fishing lines to yourself or the raft.

(e) **Kill** fish before bringing them into the raft.

(f) ***DO NOT*** eat fish with—

- Spines.
- Unpleasant odor.
- Pale, slimy gills.
- Sunken eyes.
- Flabby skin.
- Flesh that remains dented when pressed.

(g) ***DO NOT*** eat fish eggs or liver (entrails).

(h) Avoid all crustaceans above the high tide mark.

(i) Avoid cone-shaped shells **(Figure VIII-8)**.

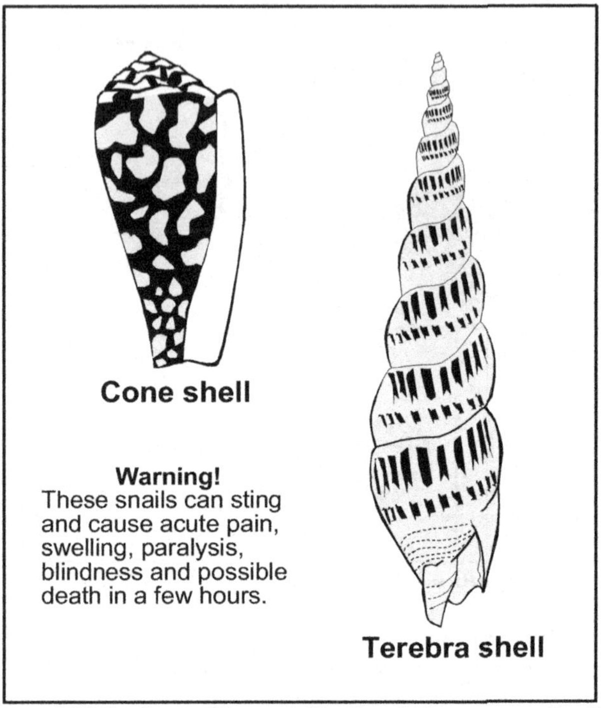

Figure VIII-8. Cone-Shaped Shells of Venomous Snails

(j) Avoid hairy insects; the hairs could cause irritation or infection.

(k) Avoid poisonous insects, for example:
- •Centipedes.
- •Scorpions.
- •Poisonous spiders.

(l) Avoid disease carrying insects, such as—
- •Flies.
- •Mosquitoes.
- •Ticks.

c. Plant Foods. ***Before using the following guide use your evasion chart to identify edible plants:***

> **Note:** If you cannot positively identify an edible plant and choose to try an unknown plant, these guidelines may help determine edibility.

(1) Selection criteria.

(a) Before testing for edibility, ensure there are enough plants to make testing worth your time and effort. Each part of a plant (roots, leaves, stems, bark, etc.) requires more than 24 hours to test. ***DO NOT*** waste time testing a plant that is not abundant.

(b) Test only **1** part of **1** plant at a time.

(c) Remember that eating large portions of plant food on an empty stomach may cause diarrhea, nausea, or cramps. ***Two*** good examples are ***green apples*** and ***wild onions***. Even after testing food and finding it safe, eat in moderation.

(2) Avoid plants with the following characteristics:

> **Note:** Using these guidelines in selecting plants for food may eliminate some edible plants; however, these guidelines will help prevent choosing potentially toxic plants.

(a) Milky sap (dandelion has milky sap but is safe to eat and easily recognizable).

(b) Spines, fine hairs, and thorns (skin irritants/contact dermatitis). ***Prickly pear*** and ***thistles*** are exceptions. ***Bracken fern fiddleheads*** also violate this guideline.

(c) Mushrooms and fungus.

(d) Umbrella shaped flowers (hemlock is eliminated).

(e) Bulbs (***only*** onions smell like onions).

(f) Grain heads with pink, purplish, or black spurs.

(g) Beans, bulbs, or seeds inside pods.

(h) Old or wilted leaves.

(i) Plants with shiny leaves.

(j) White and yellow berries. (Aggregate berries such as black and dewberries are always edible, test all others before eating.)

(k) Almond scent in woody parts and leaves.

d. Test procedures.

CAUTION: Test all parts of the plant for edibility. Some plants have both edible and inedible parts. **NEVER ASSUME** a part that proved edible when cooked is edible raw, test the part raw before eating. The same part or plant may produce varying reactions in different individuals.

(1) Test only **1** part of a plant at a time.

(2) Separate the plant into its basic components (stems, roots, buds, and flowers).

(3) Smell the food for strong acid odors. Remember, smell alone does not indicate a plant is edible or inedible.

(4) ***DO NOT*** eat 8 hours before the test and drink only purified water.

(5) During the 8 hours you abstain from eating, test for contact poisoning by placing a piece of the plant on the inside of your elbow or wrist. The sap or juice should contact the skin. Usually 15 minutes is enough time to allow for a reaction.

(6) During testing, take ***NOTHING*** by mouth **EXCEPT** purified water and the plant you are testing.

(7) Select a small portion of a single part and prepare it the way you plan to eat it.

(8) Before placing the prepared plant in your mouth, touch a small portion (a pinch) to the outer surface of your lip to test for burning or itching.

(9) If after 3 minutes there is no reaction on your lip, place the plant on your tongue and hold it for 15 minutes.

(10) If there is no reaction, thoroughly chew a pinch and hold it in your mouth for 15 minutes **(*DO NOT* SWALLOW)**. If any ill effects occur, rinse out your mouth with water.

(11) If nothing abnormal occurs, swallow the food and wait 8 hours. If **any ill effects** occur during this period, **induce** vomiting and drink a water and charcoal mixture.

(12) If no ill effects occur, eat **¼ cup** of the same plant prepared the same way. Wait another 8 hours. If no ill effects occur, the plant part as prepared is safe for eating.

CAUTION:
1. Ripe tropical fruits should be peeled and eaten raw. Softness, rather than color, is the best indicator of ripeness. Cook unripe fruits and discard seeds and skin.
2. Cook underground portions when possible to reduce bacterial contamination and ease digestion of their generally high starch content.
3. During evasion, you may not be able to cook. Concentrate your efforts on leafy green plants, ripe fruits, and above ground ripe vegetables not requiring significant preparation.

2. Food Preparation

Animal food gives the greatest food value per pound.

a. Butchering and skinning.

(1) Mammals.

(a) Remove the skin and save for other uses.

(a) One cut skinning of small game **(Figure VIII-9)**.

•Open the abdominal cavity.

•Avoid rupturing the intestines.

•Remove the intestines.

•Save inner organs (heart, liver, and kidneys) and all meaty parts of the skull, brain, tongue, and eyes.

(b) Wash when ready to use.

(c) If preserving the meat, remove it from the bones.

(d) Unused or inedible organs and entrails may be used as bait for other game.

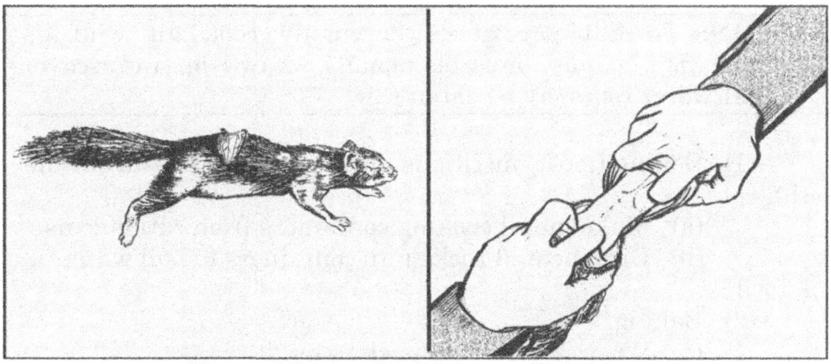

Figure VIII-9. Small Game Skinning

(2) Frogs and snakes.
(a) Skin.
(b) Discard skin, head with 2 inches of body, and internal organs.
(3) Fish.
(a) Scale (if necessary) and gut fish soon after it is caught.
(b) Insert knifepoint into anus of fish and cut open the belly.
(c) Remove entrails.
(d) Remove gills to prevent spoilage.
(4) Birds.
(a) Gut soon after killing.
(b) Protect from flies.
(c) Skin or pluck them.
(d) Skin scavengers and sea birds.
(5) Insects.
(a) Remove all hard portions such as the legs of grasshoppers or crickets. (The rest is edible.)
(b) Recommend cooking grasshopper-size insects.

CAUTION: Dead insects spoil rapidly, ***DO NOT*** save.

(6) Fruits, berries, and most nuts can be eaten raw.
b. Cooking.

CAUTION: To kill parasites, thoroughly cook all wild game, freshwater fish, clams, mussels, snails, crawfish, and scavenger birds. Saltwater fish may be eaten raw.

(1) Boiling (most nutritious method of cooking—drink the broth).
(a) Make metal cooking containers from ration cans.
(b) Drop heated rocks into containers to boil water or cook food.
(2) Baking.
(a) Wrap in leaves or pack in mud.
(b) Bury food in dirt under coals of fire.

(3) Leaching. Some nuts (acorns) must be leached to remove the bitter taste of tannin. Use one of the following leaching methods:

(a) First method:

•Soaking and pouring the water off.

•Crushing and pouring water through. Cold water should be tried first; however, boiling water is sometimes best.

•Discarding water.

(b) Second method:

•Boil, pour off water, and taste the plant.

•If bitter, repeat process until palatable.

(4) Roasting.

(a) Shake shelled nuts in a container with hot coals.

(b) Roast thinly sliced meat and insects over a candle.

3. Food Preservation

b. Keeping an animal alive.

c. Refrigerating.

(1) Long term.

(a) Food buried in snow maintains a temperature of approximately 32 degrees F.

(b) Frozen food will not decompose (freeze in meal-size portions).

(2) Short term.

(a) Food wrapped in waterproof material and placed in a stream remains cool in summer months.

(b) Earth below the surface, particularly in shady areas or along streams, is cooler than the surface.

(c) Wrap food in absorbent material such as cotton and re-wet as the water evaporates.

c. Drying and smoking removes moisture and preserves food.

(1) Use salt to improve flavor and promote drying.

(2) Cut or pound meat into thin strips.

(3) Remove fat.

(4) ***DO NOT*** use pitch woods such as fir or pine; they produce soot giving the meat an undesirable taste.

d. Protecting meat from animals and insects.
   (1) Wrapping food.
      (a) Use clean material.
      (b) Wrap pieces individually.
      (c) Ensure all corners of the wrapping are insect proof.
      (d) Wrap soft fruits and berries in leaves or moss.
   (2) Hanging meat.
      (a) Hang meat in the shade.
      (b) Cover during daylight hours to protect from insects.
   (3) Packing meat on the trail.
      (a) Wrap before flies appear in the morning.
      (b) Place meat in fabric or clothing for insulation.
      (c) Place meat inside the pack for carrying. Soft material acts as insulation helping keep the meat cool.
      (d) Carry shellfish, crabs, and shrimp in wet seaweed.

e. ***DO NOT*** store food in the shelter; it attracts unwanted animals.

Chapter IX

INDUCED CONDITIONS

(NUCLEAR, BIOLOGICAL, AND CHEMICAL CONSIDERATIONS)

1. Nuclear Conditions

> **CAUTION:** Radiation protection depends on time of exposure, distance from the source, and shielding.

a. Protection.

(1) **FIND PROTECTIVE SHELTER IMMEDIATELY!**

(2) Gather all equipment for survival (time permitting).

(3) Avoid detection and capture.

(a) Seek existing shelter that may be improved **(Figure IX-1)**.

Figure IX-1. Immediate Action Shelter

(b) If no shelter is available, dig a trench or foxhole as follows:

•Dig trench deep enough for protection, then enlarge for comfort **(Figure IX-2)**.

•Cover with available material.

Figure IX-2. Improvised Shelter

(4) Radiation shielding efficiencies **(Figure IX-3)**.

NUCLEAR EXPLOSIONS: Fall flat. Cover exposed body parts. Present minimal profile to direction of blast. ***DO NOT** look at fireball!* Remain prone until blast effects are over.

SHELTER: Pick, as soon as possible, 5 minutes unsheltered is maximum!

Priority:
(1) Cave or tunnel covered with 3 or more feet of earth.
(2) Storm/storage cellars
(3) Culverts.
(4) Basements.
(5) Abandoned stone/mud buildings.
(6) Foxhole 4 feet deep (remove topsoil within 2 feet radius of foxhole lip).

RADIATION SHIELDING EFFICIENCIES

Iron/Steel	.7 inches	**Cinder Block**	5.3 inches
Brick	2.0 inches	**Ice**	6.8 inches
Concrete	2.2 inches	**Wood (Soft)**	8.8 inches
Earth	3.3 inches	**Snow**	20.3 inches

One thickness reduces received radiation dose by 1/2. Additional thickness added to any amount of thickness reduces received radiation dose by 1/2.

SHELTER SURVIVAL: Keep contaminated materials out of shelter.
Good Weather: Bury contaminated clothing outside of shelter (recover later).
Bad Weather: Shake strongly or beat with branches. Rinse and /or shake wet clothing. ***DO NOT** wring out!*

PERSONAL HYGIENE: Wash entire body with soap and *any* water; give close attention to fingernails and hairy parts.
No Water: Wipe all exposed skin surfaces with clean cloth or uncontaminated soil. Fallout/dusty conditions--keep entire body covered. Keep handkerchief/cloth over mouth and nose. Improvise goggles. ***DO NOT** smoke!*

DAILY RADIATION TIME TABLE for NO RATE METER

4-6	Complete isolation	**9-12**	2-4 hours exposure per day
3-7	Brief exposure (30 minutes maximum)	**13**	Normal movement.
8	Brief exposure (1 hour maximum)		

Figure IX-3. Radiation Shielding Efficiencies

(5) Leave contaminated equipment and clothing near shelter for retrieval after radioactive decay.

(6) Lie down, keep warm, sleep, and rest.

b. Substance:

(1) Water. Allow no more than 30 minutes exposure on **3d** day for water procurement.

(a) Water sources (in order of preference):

- Springs, wells, or underground sources are **safest**.
- Water in pipes/containers in abandoned buildings.
- Snow (**6** or **more inches below** the surface during the fallout).

•Streams and rivers (filtered before drinking).
•Lakes, ponds, pools, etc.
•Water from below the surface (***DO NOT*** stir up the water).
•Use a seep well.

(b) Water preparation **(Figures IX-4 and IX-5)**.
•Filtering through earth removes 99 percent of radioactivity.
•Purify all water sources.

(2) Food.

(a) Processed foods (canned or packaged) are preferred; wash and wipe containers before use.

(b) Animal foods.
•Avoid animals that appear to be sick or dying.
•Skin carefully to avoid contaminating the meat.
•Before cooking, cut meat away from the bone, leaving at least 1/8 inch of meat on the bone.
•Discard all internal organs.
•Cook all meat until **very well** done.

(c) Avoid.
•Aquatic food sources (use only in extreme emergencies because of high concentration of radiation).
•Shells of all eggs (contents will be safe to eat).
•Milk from animals.

(d) Plant foods (in order of preference).
•Plants whose edible portions grow underground (for example, potatoes, turnips, carrots, etc.). Wash and remove skin.
•Edible portions growing above ground that can be washed and peeled or skinned (bananas, apples, etc.).
•Smooth skinned vegetables, fruits, or above ground plants that are not easily peeled or washed.

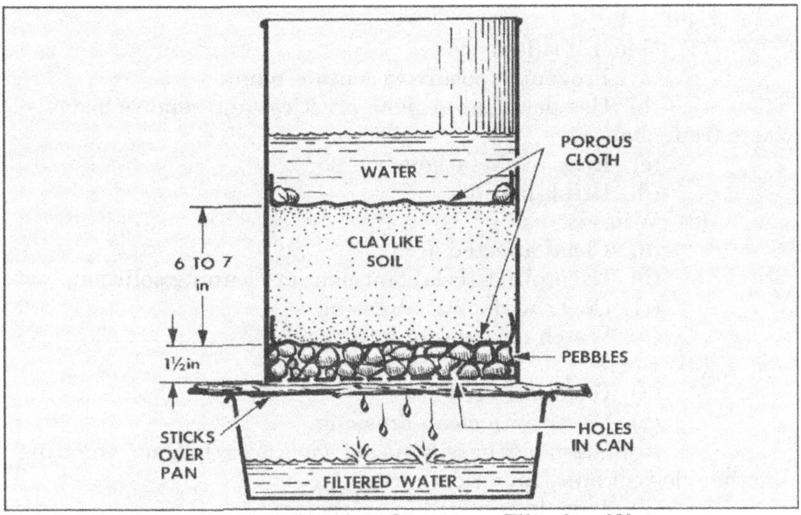

Figure IX-4. Filtration Systems, Filtering Water

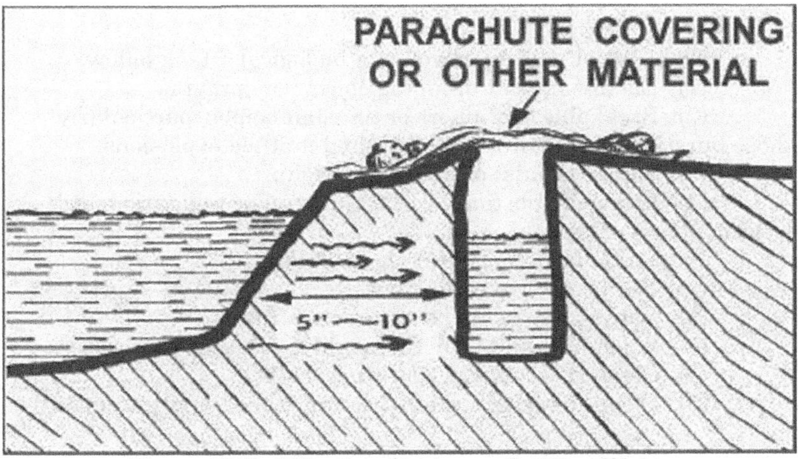

Figure IX-5. Filtration Systems, Settling Water

c. Self-aid:

(1) General rules:

(a) Prevent exposure to contaminants.

(b) Use personal hygiene practices and remove body waste from shelter.

(c) Rest, avoid fatigue.

(d) Drink liquids.

(2) Wounds.

(a) Clean affected area.

(b) Use antibacterial ointment or cleaning solution.

(c) Cover with clean dressing.

(d) Watch for signs of infection.

(3) Burns.

(a) Clean affected area.

(b) Cover with clean dressing.

(4) Radiation sickness (nausea, weakness, fatigue, vomiting, diarrhea, loss of hair, radiation burns).

(a) Time is required to overcome.

(b) Rest.

(c) Drink fluids.

(d) Maintain food intake.

(e) Prevent additional exposure.

2. Biological Conditions

a. Clues which may alert you to a biological attack follow:

(1) Enemy aircraft dropping objects or spraying.

(2) Breakable containers or unusual bombs, particularly those bursting with little or no blast, and muffled explosions.

(3) Smoke or mist of unknown origin.

(4) Unusual substances on the ground or vegetation; sick looking plants or crops.

b. Protection from biological agents follow:

(1) Use protective equipment.

(2) Bathe as soon as the situation permits.

(3) Wash hair and body thoroughly with soap and water.

(4) Clean thoroughly under fingernails.

(5) Clean teeth, gums, tongue, and roof of mouth frequently.

c. Survival tips for biological conditions follow:

(1) Keep your body and living area clean.

(2) Stay alert for clues of biological attack.

(3) Keep nose, mouth, and skin covered.

(4) Keep food and water protected. Bottled or canned foods are safe if sealed. If in doubt, boil food and water for 10 minutes.

(5) Construct shelter in a clear area, away from vegetation, with entrance 90 degrees to the prevailing wind.

(6) If traveling, travel crosswind or upwind (taking advantage of terrain to stay away from depressions).

3. Chemical Conditions

a. Detecting.

(1) Smell. Many agents have little or no odor.

(2) Sight. Many agents are colorless:

(a) Color. Yellow, orange, or red smoke or mist.

(b) Liquid. Oily, dark patches on leaves, ground, etc.

(c) Gas. Some agents appear as a mist immediately after shell burst.

(d) Solid. Most solid state agents have some color.

(3) Sound. Muffled explosions are possible indications of chemical agent bombs.

(4) Feel. Irritation to the nose, eyes, or skin and/or moisture on the skin are danger signs.

(5) Taste. Strange taste in food or water indicates contamination.

(6) General indications. Tears, difficult breathing, choking, itching, coughing, dizziness.

(7) Wildlife. Presence of sick or dying animals.

b. Protection against chemical agents follows:

(1) Use protective equipment.

(2) Avoid contaminated areas.

(a) Exit contaminated area by moving crosswind.

(b) Select routes on high ground.

(c) Avoid cellars, ditches, trenches, gullies, valleys, etc.

(d) Avoid woods, tall grasses, and bushes as they tend to hold chemical agent vapors.

(e) Decontaminate body and equipment as soon as possible by—

- •Removing. Pinch-blotting.
- •Neutralizing. Warm water.
- •Destroying. Burying.

c. Self-aid in chemically contaminated areas.

(1) If a chemical defense ensemble is available—

(a) Use all protective equipment.

(b) Follow antidote directions when needed.

(2) If a chemical defense ensemble is not available—

(a) Remove or tear away contaminated clothing.

(b) Rinse contaminated areas with water.

(c) Improvise a breathing filter using materials available (T-shirt, handkerchief, fabric, etc.).

d. Tips for the survivor:

(1) ***DO NOT*** use wood from a contaminated area for fire.

(2) Look for signs of chemical agents around water sources before procurement (oil spots, foreign odors, dead fish, or animals).

(3) Keep food and water protected.

(4) ***DO NOT*** use plants for food or water in contaminated areas.

Appendix A
THE WILL TO SURVIVE

ARTICLE VI CODE OF CONDUCT

I will never forget that I am an American fighting for freedom, responsible for my actions, and dedicated to the principles which made my country free. I will trust in my God and in the United States of America.

1. Psychology of Survival

a. Preparation—

(1) Know your capabilities and limitations.

(2) Keep a positive attitude.

(3) Develop a realistic plan.

(4) Anticipate fears.

(5) Combat psychological stress by—

(a) Recognizing and anticipating existing *stressors* (injury, death, fatigue, illness, environment, hunger, isolation).

(b) Attributing normal reactions to existing *stressors* (fear, anxiety, guilt, boredom, depression, anger).

(c) Identifying signals of distress created by *stressors* (indecision, withdrawal, forgetfulness, carelessness, and propensity to make mistakes).

b. Strengthen your will to survive with—

(1) The Code of Conduct.

(2) Pledge of Allegiance.

(3) Faith in America.

(4) Patriotic songs.

(5) Thoughts of return to family and friends.

c. Group dynamics of survival include—

(1) Leadership, good organization, and cohesiveness promote high morale:

(a) Preventing panic.

(b) Creating strength and trust in one another.

(c) Favoring persistency in overcoming failure.

(d) Facilitating formulation of group goals.

(2) Taking care of your buddy.

(3) Working as a team.

(4) Reassuring and encouraging each other.

(5) Influencing factors are—
    (a) Enforcing the chain of command.
    (b) Organizing according to individual capabilities.
    (c) Accepting suggestions and criticism.

2. Spiritual Considerations

a. Collect your thoughts and emotions.
b. Identify your personal beliefs.
c. Use self-control.
d. Meditate.
e. Remember past inner sources to help you overcome adversity.
f. Pray for your God's help, strength, wisdom, and rescue.
    (1) Talk to your God.
    (2) Give thanks that God is with you.
    (3) Ask for God's help.
    (4) Pray for protection and a positive outcome.
g. Remember scripture, verses, or hymns; repeat them to yourself and to your God.
h. Worship without aid of written scripture, clergy, or others.
i. Forgive—
    (1) Yourself for what you have done or said that was wrong.
    (2) Those who have failed you.
j. Praise God and give thanks because—
    (1) God is bigger than your circumstances.
    (2) God will see you through (no matter what happens).
    (3) Hope comes from a belief in heaven and/or an after-life.
k. Trust.
    (1) Faith and trust in your God.
    (2) Love for family and self.
    (3) Never lose hope.
    (4) Never give up.
l. With other survivors—
    (1) Identify or appoint a religious lay leader.
    (2) Discuss what is important to you.
    (3) Share scriptures and songs.
    (4) Pray for each other.
    (5) Try to have worship services.

(6) Write down scriptures and songs that you remember.

(7) Encourage each other while waiting for rescue, remember—

(a) Your God loves you.

(b) Praise your God.

Appendix B
Publication Information

1. Scope

This UNCLASSIFIED multiservice tactics, techniques, and procedures publication is designed to assist Service members in a survival situation regardless of geographic location.

2. Purpose

This publication provides Service members a quick reference, weatherproof, pocket-sized guide on basic survival, evasion, and recovery information.

3. Application

The target audience for this publication is any Service member requiring basic survival, evasion, and recovery information.

4. Implementation Plan

Participating Service command offices of primary responsibility (OPRs) will review this publication, validate the information, and reference and incorporate it in Service and command manuals, regulations, and curricula as follows:

Army. The Army will incorporate the procedures in this publication in US Army training and doctrinal publications as directed by the commander, US Army Training and Doctrine Command (TRADOC). Distribution is in accordance with the Department of Army Form 12-99-R.

Marine Corps. The Marine Corps will incorporate the procedures in this publication in US Marine Corps training and doctrinal publications as directed by the commanding general, US Marine Corps Combat Development Command (MCCDC). Distribution is in accordance with Marine Corps Publication Distribution System.

Navy. The Navy will incorporate these procedures in US Navy doctrinal and training publications as directed by the commander, Navy Warfare Development Command (NWDC). Distribution is in accordance with Military Standard Requisitioning and Issue Procedures Desk Guide and Navy Standing Operating Procedures Publication 409.

Air Force. Air Force units will validate and incorporate appropriate procedures in accordance with applicable governing directives. Distribution is in accordance with Air Force Instructions 33-360.

5. User Information

a. The TRADOC-MCCDC-NWDC-AFDC Air Land Sea Application (ALSA) Center developed this publication with the joint participation of the approving Service commands. ALSA will review and update this publication as necessary.

b. This publication reflects current joint and Service doctrine, command and control (C2) organizations, facilities, personnel, responsibilities, and procedures. Changes in Service protocol, appropriately reflected in joint and Service publications, will likewise be incorporated in revisions to this document.

c. We encourage recommended changes for improving this publication. Key your comments to the specific page and paragraph and provide a rationale for each recommendation. Send comments and recommendation directly to—

Army

Commander
US Army Training and Doctrine Command
ATTN: ATDO-A
Fort Monroe VA 23651-5000
DSN 680-3153 COMM (757) 727-3153

Marine Corps

Commanding General
US Marine Corps Combat Development Command
ATTN: C42
3300 Russell Road
Quantico VA 22134-5001
DSN 278-6234 COMM (703) 784-6234

Navy

Commander
Navy Warfare Development Command (Det Norfolk)
ATTN: ALSA Liaison Officer
1530 Gilbert Street
Norfolk VA 23511-2785
DSN 262-2782 COMM (757) 322-2782
E-mail: ndcjoint@nctamslant.navy.mil

Air Force

HQ Air Force Doctrine Center
ATTN: DJ
216 Sweeney Boulevard Suite 109
Langley AFB VA 23665-2722
DSN 754-8091 COMM (757) 764-8091
E-mail Address: afdc.dj@langley.af.mil

ALSA

ALSA Center
ATTN: Director
114 Andrews Street
Langley AFB, VA 23665-2785
DSN 575-0902 COMM (757) 225-0902
E-mail: alsadirector@langley.af.mil

Made in United States
North Haven, CT
04 November 2026

21774370R00137